D0078252

INTRODUCTION TO BIOMEDICAL INSTRUMENTATION

The Technology of Patient Care

This book is designed to introduce the reader to the fundamental information necessary for work in the clinical setting, supporting the technology used in patient care. Beginning biomedical equipment technologists can use this book to obtain a working vocabulary and elementary knowledge of the industry. Content is presented through the inclusion of a wide variety of medical instrumentation, with an emphasis on generic devices and classifications; individual manufacturers are explained only when the market is dominated by a particular unit. This book is designed for the reader with a fundamental understanding of anatomy, physiology, and medical terminology appropriate for their role in the health care field and assumes the reader's understanding of electronic concepts, including voltage, current, resistance, impedance, analog and digital signals, and sensors. The material covered in this book will assist the reader in the development of his or her role as a knowledgeable and effective member of the patient care team.

Barbara L. Christe is Associate Professor and Program Director of Biomedical Engineering Technology at Indiana University Purdue University Indianapolis.

INTRODUCTION TO BIOMEDICAL INSTRUMENTATION

The Technology of Patient Care

Barbara L. Christe

Indiana University Purdue University Indianapolis

CAMBRIDGE UNIVERSITY PRESS
Cambridge, New York, Melbourne, Madrid, Cape Town, Singapore, São Paulo, Delhi

Cambridge University Press
32 Avenue of the Americas, New York, NY 10013-2473, USA

www.cambridge.org
Information on this title: www.cambridge.org/9780521515122

© Cambridge University Press 2009

This publication is in copyright. Subject to statutory exception
and to the provisions of relevant collective licensing agreements,
no reproduction of any part may take place without the written
permission of Cambridge University Press.

First published 2009

Printed in the United States of America

A catalog record for this publication is available from the British Library.

Library of Congress Cataloging in Publication data
Christe, Barbara L., 1962–
Introduction to biomedical instrumentation : the technology of patient care / Barbara L. Christe.
p. ; cm.
Includes bibliographical references and index.
ISBN 978-0-521-51512-2 (hbk.)
1. Medical instruments and apparatus. 2. Biomedical engineering. 3. Medical electronics. I. Title. [DNLM: 1. Biomedical Technology – instrumentation. 2. Biomedical Engineering – instrumentation. 3. Electronics, Medical – instrumentation. 4. Patient Care – instrumentation. W 26 C554i 2009]
R856.C52 2009
610.28–dc22 2008039462

ISBN 978-0-521-51512-2 hardback

Every effort has been made in preparing this book to provide accurate and up-to-date information that is in accord with accepted standards and practice at the time of publication. Although case histories are drawn from actual cases, every effort has been made to disguise the identities of the individuals involved. Nevertheless, the authors, editors, and publishers can make no warranties that the information contained herein is totally free from error, not least because clinical standards are constantly changing through research and regulation. The authors, editors, and publishers therefore disclaim all liability for direct or consequential damages resulting from the use of material contained in this book. Readers are strongly advised to pay careful attention to information provided by the manufacturer of any drugs or equipment that they plan to use.

Cambridge University Press has no responsibility for the persistence or accuracy of URLs for external or third-party Internet Web sites referred to in this publication and does not guarantee that any content on such Web sites is, or will remain, accurate or appropriate. Information regarding prices, travel timetables, and other factual information given in this work are correct at the time of first printing, but Cambridge University Press does not guarantee the accuracy of such information thereafter.

Contents

Preface

This book serves readers who would like to explore medical equipment that is used in the clinical setting. It offers an overview of the fundamental information necessary for work in the field. The material is designed to provide a working vocabulary and *elementary* knowledge of the medical equipment involved in the treatment of patients. Readers are encouraged to become knowledgeable and effective members of the patient care team, building on the information in this book as a foundation for further study.

Readers should have a fundamental understanding of *anatomy, physiology, and medical terminology* appropriate for their role in the health care field. Readers are assumed to have a fundamental knowledge of *basic electronics* concepts including voltage, current, resistance, impedance, analog and digital signals, and sensors. Readers without this background may have to explore terms and concepts referenced in the text.

There is a vital connection between technology and the care of patients. In many cases, health care workers depend on technology to administer care or treatment or to make a diagnosis. This book helps readers understand how technology is tightly woven into patient care. The role of technical support for the medical team is, therefore, essential in the delivery of effective medical care.

The section of each chapter entitled "For Further Exploration" encourages readers to use the Internet to obtain in-depth information about a topic. Questions are designed to push the reader to integrate concepts using external sources. Answers are not specifically available within the chapters. While Wikipedia (http://www.wikipedia.com) may not be an academically authoritative source, it is often an excellent starting point for research. Research exercises encourage one of the most important skills of a successful biomedical equipment technician (BMET) – investigation of topics that are not well understood. In the clinical setting, it is impossible to be an expert about all technology and aspects of patient care. The ability to effectively search for information is vital.

All photographs (unless otherwise noted) are by Valerie Shiver.

Special appreciation is extended to my collegiate mentors, Dean Jeutter and Joe Bronzino. I am in debt to my colleagues in higher education, Steve Yelton, Roger Bowles, Elaine Cooney, and Ken Reid, as well as those who have supported me in the clinical setting, including Dave Francouer, Karen Waninger, Kelly Vandewalker, Steve Erdosy, and Bob Pennington. Without their encouragement, this book could not have been written. Lastly, to my parents and children, I am deeply grateful for their love and understanding.

Those who support the technology used in patient care are a dedicated and selfless part of the workforce. They are part of the medical team and have excellent technical skills. Most importantly, BMETs work closely with staff to ensure safe and effective patient care. May this book be the beginning of a transformation that increases career awareness, improves enrollment in training programs, and expands the recognition BMETs deserve.

INTRODUCTION TO BIOMEDICAL INSTRUMENTATION

The Technology of Patient Care

1

BMET as a career

LEARNING OBJECTIVES

1. describe the role of a BMET
2. list and describe potential employers of a BMET
3. characterize field service representatives
4. list and describe the many job functions of a BMET
5. list and characterize the certification requirements
6. list and describe related professional societies and journals

What is the name of this career?

There are many definitions for what the letters BMET stand for – biomedical equipment technician, biomedical electronics technology, medical maintenance, biomedical engineering technologist, biomedical engineer, medical engineer, medical equipment repair technician, and many more. In general, it is the title for someone who works in the clinical setting and supports the equipment involved in patient care.

At different hospitals, staff may have a wide variety of titles. Staff may be called the "biomeds," "clinical engineers," or the "equipment guys." In some hospitals, BMETs may be responsible for everything from printers to computers to DVD players in the rooms of patients. In some hospitals, BMETs work for the maintenance departments and wear janitor's jumpsuits. Other hospitals hire a wide range of technical staff who wear lab coats, monogrammed polo shirts, or dress shirts.

Because there is so little uniformity, it can be difficult for the career field to get the recognition it deserves. Some basic facts are true for most BMETs.

What do BMETs do?

Most BMETs perform several main categories of job function. In general, BMETs are responsible for the support of the technology used in health care. This support assures the safe use of equipment and the best possible patient care. BMETs work closely with medical staff to make sure technology is used safely and effectively.

Ultimately, the *customer* of BMET services is the patient, although many times the patient and BMET are not in the same place at the same time. Most experienced BMETs define the best BMET as one who thinks of each patient as a relative or loved one. The care and attention one would expect under these circumstances should drive a BMET's job performance.

Who employs BMETs?

Generally, there are three groups of employers of BMETs: hospitals, outside service providers, and the manufacturers themselves. Those who work for the hospital directly or an outside service organization (OSA) or independent service organization (ISO) may appear the same to clinical staff. Some employment issues (benefits, etc.) could be different, but the work-related duties are likely to be similar. Often, when employees work for a manufacturer they are identified as field service representatives or FSRs.

Hospital-employed BMETs generally have the following responsibilities:

- **Equipment repair and troubleshooting** – BMETs fix equipment that is not functioning as expected. This repair may or may not be done "in the shop." BMETs may need to retrieve equipment that has a "broke" sign attached to it, or they may be called to the operating room during a case. Figure 1.1 shows a BMET working on a physiological monitor at her workstation in the clinical engineering shop of the hospital.
- **Preventative maintenance (PM)** – BMETs routinely verify the performance of almost all equipment. This involves evaluating the performance of every aspect of a device and checking or replacing parts to ensure consistent, dependable service. PM may include conducting calibrations and

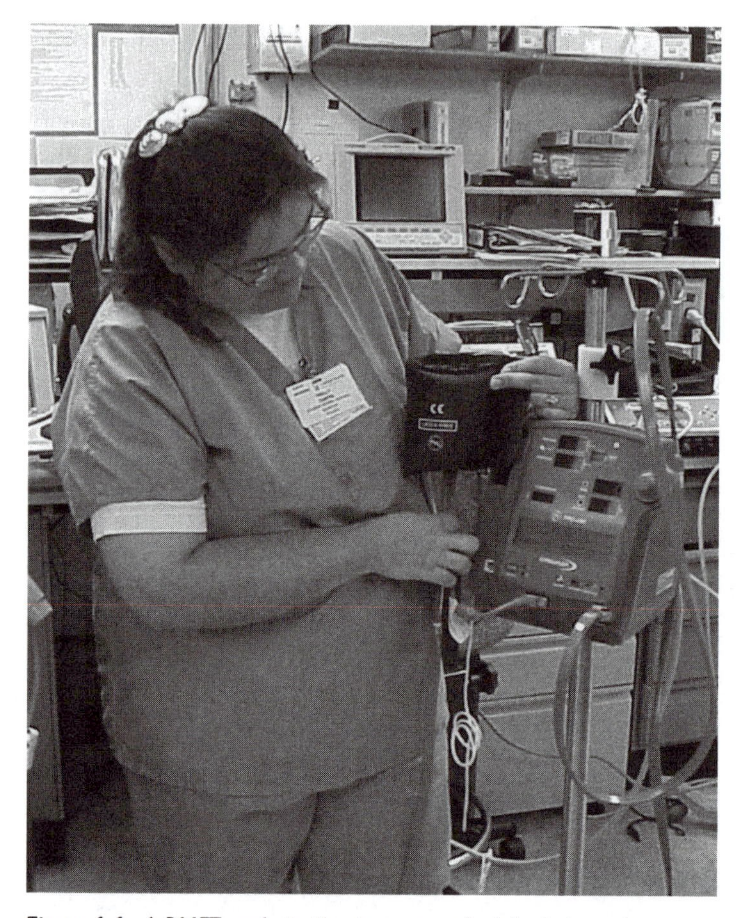

Figure 1.1. A BMET works in the shop on a physiological monitor.

safety checks as well as removing the "white dust" that comes from bed linens, a task that occupies a great deal of the fledgling BMET's time. Performing preventive maintenance is a great way to learn about all the features of an instrument, and the experience can assist a BMET in future troubleshooting. This type of activity is sometimes called *performance assurance*. Figure 1.2 shows a BMET performing PM on a ventilator.

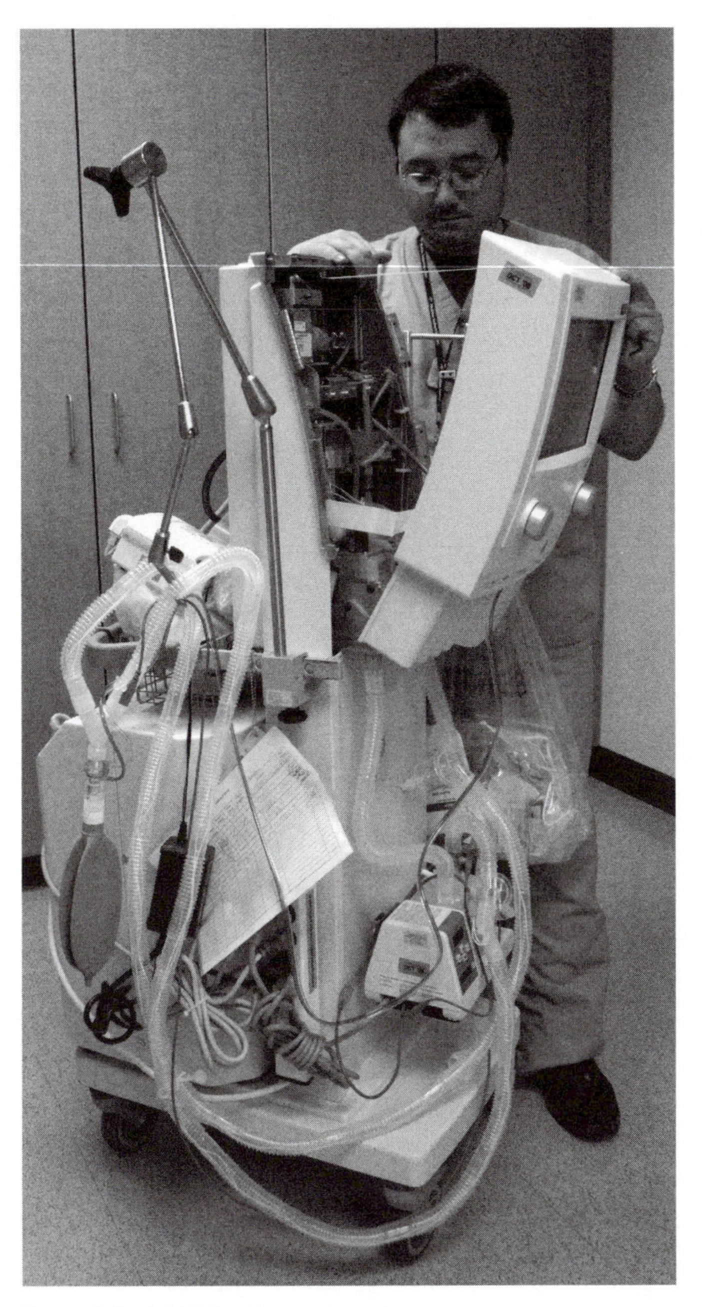

Figure 1.2. A BMET calibrates a ventilator.

- **Staff support** – BMETs provide both formal and informal equipment instruction to many groups including the users of equipment and other BMETs. BMETs may arrange and lead an *in-service meeting* during a staff meeting to introduce a new device and train staff. BMETs also work one-on-one with a staff member. Excellent customer service is vital to effective job performance.
- **Pre-purchase evaluation** – As new equipment (new models or entirely new devices) is considered for purchase, many BMETs are involved in the selection decisions, usually working very closely with the medical staff. As medical technology becomes interwoven with other medical equipment in the hospital, the cross-departmental interactions often fall to BMETs.
- **Incident investigation** – When there are problems with equipment, experienced BMETs are often part of the team that evaluates issues surrounding a malfunction.
- **Incoming testing** – When new devices arrive at the hospital, BMETs must verify that every aspect of every piece of equipment functions properly.
- **Adaptations/modifications** – BMETs are occasionally asked to modify equipment to better medically serve clinical staff as well as better serve a patient with restrictions or limitations.
- **Departmental development/training classes** – Departments have meetings and other activities that must be documented. Documentation of activities is a required departmental activity. Accreditation bodies have a policy that basically concludes: "if it is not written down, it did not happen." In addition, BMETs are often expected to participate in additional training, which is usually device specific and often offered by the manufacturer.

- **Updates** – When manufacturers change or update equipment (for example, software) the BMET installs or makes the necessary changes.
- **Safety board** – BMETs help to set policies and investigate problems, especially regarding hospital process efficiency and staff training. In addition, plans for emergencies include medical equipment, and BMETs contribute to these disaster plans.

Non-hospital-employed BMETs take on a number of the previous responsibilities in addition to some of the following functions:

- **Telephone support** – Some BMETs answer phone lines to assist users of equipment as well as technicians who are attempting to make a repair.
- **Sales** – Some BMETs work for a manufacturer, outside service organization, or repair depot as a salesperson.
- **New equipment design** – Some BMETs work for a manufacturer and design new devices.

In general, most hospital-based BMETs work "hospital" hours: 7 A.M. to 3:30 P.M. (or something like that) Monday through Friday. While some institutions do have shifts on weekends, most shops do not staff nights or weekends. Policies for "on-call" coverage vary, although most hospitals easily deal with problems "off hours" with minimal weekend and night trips back to work.

Table 1.1 shows the standardized definitions for people who support medical equipment technology. Levels have been identified based on years of experience, and this table also explains the higher education requirements that are most common (although not absolutely required).

TABLE 1.1. *AAMI job descriptions*

BMET I – An entry-level or junior biomedical equipment technician (BMET). Works under close supervision. Performs skilled work on preventive maintenance, repair, safety testing, and recording functional test data. Not certified. Usually has less than four years of experience.
BMET II – A BMET who usually has a two-year degree or higher. Has good knowledge of schematics and works independently on repairs, safety testing, and preventive maintenance (PM). Maintains records, writes reports, and coordinates outside repairs. Average experience is eight years.
BMET III – A highly experienced or specialized BMET usually having an AS (two-year) degree or higher. Has substantial experience and may be certified (CBET). Does highly skilled work of considerable difficulty. Has comprehensive knowledge of practices, procedures, and types of equipment. Average experience is 12 years.
Equipment Specialist – A highly specialized BMET having special training or equivalent experience in lab equipment (LES) or radiology equipment (RES). Usually has an AS (two-year) degree or higher. Performs highly skilled work of considerable difficulty and may hold certification as CLES or CRES.
BMET Supervisor – A BMET who supervises others. Has a significant amount of training, education, or equivalent experience. Most have a BS (four-year) degree or higher. Schedules and assigns work to subordinates, but also continues to do highly skilled repairs. Has comprehensive knowledge of practices, procedures, and types of equipment. Average experience is 13 years.
Clinical Engineer – A graduate engineer holding a BS, MS, or PhD. Performs engineering-level work of considerable difficulty. Has the ability to modify devices and do analysis of devices and systems.
Clinical Engineering Supervisor – A clinical engineer who supervises BMET/peer/subordinate clinical engineers; may also supervise equipment specialists. Usually degreed engineer at BA, MS, or PhD level. Average experience is 21 years.
Director/Department Manager – Most are educated or experienced as clinical engineers or BMETs, but others may be trained in administration or business or have extensive health care supervisory experience. Most have a significant amount of technical or management experience and the skills to select high-tech equipment and acquire, maintain, and repair equipment. Supervises BMETs, clinical engineers, and support personnel. May also be the chief technology officer or vice president for health care technology.
IT Technologist/Technician – An IT technologist/technician manages projects in the areas of system administration, software development, and network security and provides direct technical support in at least one of these areas.

Source: Biomedical Instrumentation and Technology, January 2008, p. 26. Reprinted with permission from *Biomedical Instrumentation & Technology*, a peer-reviewed journal by the Association for the Advancement of Medical Instrumentation. Visit www.aami.org to learn more about AAMI or to view BI&T's current issue.

Field service representatives

Field service representatives (FSRs) are generally employed by the manufacturer of a medical device or technology. This person represents the company by servicing or supporting (training, for example) a particular device or group of devices at the clinical site. Sometimes these BMETs are called field service engineers, equipment specialists, or customer engineers.

In general, field service representatives perform many of the job functions of general BMETs. The proportion of the time spent on the various facets of the job shifts when FSRs focus on one type or group of equipment. In addition, some FSRs are very specialized as trainers or do mainly repairs and therefore have a very narrow range of duties.

Generally, FSRs are commonly used in such areas as radiology (imaging), clinical laboratory, anesthesia, LASERs, and operating room equipment, to name a few. Most common is imaging and clinical lab since they involve very complex, very expensive equipment that requires in-depth (weeks or months of) training. It is a significant financial and personnel commitment for individual institutions to train people to support a single device (or a few) that one hospital owns. By spreading the technical skills of an FSR over several hospitals, support expenses to the institution may be less. Or, contracting for service may be the only option an institution has to provide a skilled technician who can support the device (irrespective of cost).

Most field service representatives work under a service contract purchased by a clinical facility. Some manufacturers require that service only be performed by their own FSRs. In addition, service contracts can be efficient for the institution because a highly trained person will respond quickly. This may be especially true when downtime of a particular device adversely affects patient care.

There are generally two types of service contracts. A full service contract specifies that an FSR will respond within a certain period of time and repair the equipment with no additional costs. The exact financial arrangement and details are usually negotiated. Another category of service contract provides less comprehensive service. For example, the contract may be "time and materials" – this allows a hospital to have access to an FSR and still pay them an hourly rate (usually with a minimum number of hours) as well as the cost of the parts. Some service contracts allow in-house BMETs to look at the device to try to identify any simple issues; this first-response technique may also include phone technical support for the in-house BMET.

Many FSRs travel between clinical sites. In some territories, this travel may be minutes, in larger states, the amount of time on the road can be hours. The territory that an FSR covers may also impact the number of nights that are not spent at home. Many FSRs work out of their car, stocking parts in the trunk and completing paperwork in hotel rooms or at home. Many FSRs are on call 24 x 7 and may be required to stay at a site until a repair is complete. Overtime is relatively common, and their schedule may not be very predictable. To compensate BMETs for these challenges, the salaries offered to field service reps are usually very good (and often higher than hospital-based BMETs). Benefits often include a company car and other travel expenses. There may be incentives and bonuses available. Long-term salary surveys show large salary improvements for those BMETs who specialize in areas such as imaging and the clinical lab.

Many BMET students express an interest in specialization, mainly because of the high salary potential. It may be difficult to secure a position as an FSR without some BMET experience. An FSR is the only person at a site to solve a problem. Very few companies will consider hiring a person who has never worked as a BMET and never been in the clinical setting (except perhaps during their internship) to shoulder the heavy responsibility of expensive equipment and corporate reputation. Willingness to relocate may also be required. Lastly, in addition to excellent technical skills, FSRs must have excellent customer service skills. The ability to communicate with the medical staff, especially when there are difficulties and delays, can be vital to the relationship between the manufacturer and the clinical site. Personality and professionalism will be absolutely required to locate a position as an FSR.

Is there a national license or certification?

Unlike nursing and other medical professions, there is no licensure required to be a BMET. Optional national and international certifications are available; however, employers vary as to the emphasis certification receives.

The most common certification is offered by the International Certification Commission. The process and procedures are overseen by AAMI (Association for the Advancement of Medical Instrumentation). This group offers three types of certification, which are indentified in Table 1.2. The first step to obtain certification is to take a national exam (there is a separate exam for each of the three types of certification). Applicants with an associate's degree in BMET can take the CBET exam as a **candidate.** Applicants who have this 2-year degree and 2 years of work experience in the field can take the exam for **full certification** status.

TABLE 1.2. *Types of ICC certification*

Type of certification	Position
CBET	Biomedical equipment technicians
CRES	Radiology equipment specialists
CLES	Laboratory equipment specialists

Requirements for each of the other types of exams are available by visiting the AAMI Web site. There is a fee to take this exam, but most employers will reimburse the cost if you are successfully certified.

Details about the exam are located at http://www.aami.org/certification/about.html.

There are five sections of the CBET exam:

- Anatomy and Physiology
- Safety in the Health Care Facility
- Fundamentals of Electronics
- Medical Equipment Function and Operation
- Medical Equipment Problem Solving

Most successful candidates study before the test.

Less common is certification obtained through the Electronics Technicians Association-International (ETA-I), which offers both a student and a professional certification or The International Society of Certified Electronics Technicians (ISCET). Both organizations offer Certified Electronics Technician (CET) exams in the biomedical field.

Certification can improve pay rates but generally not a great deal. It is not required for employment by any government body. Very few employers require certification for employment but many employers recommend it.

What regulatory agencies govern the work of BMETs?

Numerous governing bodies and associations guide the use of equipment in medical care. Many groups do not regulate using laws but rather offer guidelines and validation of compliance with standards of best practice. All hospitals are legally regulated by the board of health for a specific municipality. However, most specific guidelines do not come from state and local legislatures. The most prominent agency is The Joint Commission (previously known as The Joint Commission on Accreditation of Healthcare Organizations, JCAHO). It is an independent, not-for-profit organization that does not specifically "regulate" hospitals but offers voluntary accreditation. With this accreditation, a hospital is eligible for Medicaid and Medicare payments. While technically optional, almost all hospitals are inspected by The Joint Commission in order to be reimbursed for patients covered by Medicare and Medicaid. The Joint Commission guides many hospital activities, not just the support of technology. Other agencies and associations include the National Fire Protection Association (NFPA), Compressed Gas Association (CGA), College of American Pathology (CAP), Occupational Health and Safety Association (OSHA), the Laser Safety Institute (LSA), and the Association for the Advancement of Medical Instrumentation.

What are some ways that BMETs stay connected?

- Subscribe to *24x7*, a *free* magazine designed for BMETs. Visit http://www.24x7mag.com to subscribe.
- Subscribe to *Medical Dealer*, a *free* magazine that contains articles for biomedical technicians. To subscribe, visit the Web site at www.mdpublishing.com.

- Join **Biomedtalk-L.** Biomedtalk is an email listserv that allows BMETs to communicate about a wide variety of topics, some very technical, some very humorous. There is a small fee to subscribe. Visit www.bmetsonline.org.
- Join the **Association for the Advancement of Medical Instrumentation**. AAMI has several excellent publications as well as a large annual conference. For students enrolled in at least 12 credit hours, the cost is extremely low. The Web site for the association is http://www.aami.org.
- Join your **local society**. Many areas of the United States have organizations. A list of regional groups is available on the AAMI Web site http://www.aami.org/resources/links/biomed.html.
- **META** (Medical Equipment and Technology Association) is a newer BMET association that has some good resources. Its site is http://www.mymeta.org/.

Note that BMET is sometimes confused with biomedical engineering (BME), which is not very closely related to the work of BMETs. Most biomedical engineers are focused more on the research than on the support of existing devices and technologies. Many biomedical engineers are examining issues at a cellular level and do not have a foundation in electronics. You can visit the Biomedical Engineering Society Web site at http://www.bmes.org to see the differences clearly. A point of great confusion occurs when hospitals and societies label BMETs as biomedical engineers.

STUDY QUESTIONS

1. Write a brief want ad you might see for a BMET entry-level position. Include typical duties and qualifications.

2. Where does a BMET usually work? Who are typical employers?
3. What is an in-service meeting? Who would attend?
4. What is done during a PM? Why are they beneficial?
5. Describe a typical day for hospital-based, entry-level BMETs. What would the BMETs wear? How might they spend their time? Where in the hospital would they be doing these activities?
6. Make a list of the advantages and disadvantages of field service work.
7. Define and describe certification. Is it voluntary? Why might a BMET become certified?
8. Being inspected by The Joint Commission is technically optional but why is it important to so many facilities?
9. Make a list of some of the groups that you might consider joining as part of your career. List some of the benefits of joining associations and societies.

FOR FURTHER EXPLORATION

1. The lack of career awareness can be a significant hurdle. Visit the government's *Occupational Outlook Handbook* (from the U.S. Department of Labor) and search for "medical equipment repairers." Summarize the information presented. Does the information look accurate for the positions in your area?
2. Visit AAMI's BMET career Web site at http://www.aami.org/resources/BMET. Summarize the information presented.
3. Watch a video made about the BMET field. It is located at http://www.learnmoreindiana.org/careers/exploring/Pages/CareerProfiles.aspx?VID=7&SOC=49906200&LID=0&RFP=1&RBP=557. Summarize the information presented.

4. In what year was nursing founded? Now, consider that technology was not used in health care until the 1970s. BMETs did not exist prior to the introduction of technology into patient care. What kind of impact has the relatively short history of biomedical instrumentation had on the prestige and recognition of BMETs within the hospital? For example, can you understand how space allocation in a hospital is influenced by "who got there first"? Use the Internet to collect references to substantiate your answers.
5. A significant rise in BMET career awareness occurred with an article written by Ralph Nader in *Ladies Home Journal* in March 1971; it claimed that there were a large number of hospital electrocutions each year. Search the Internet for this infamous article and summarize it. Evaluate the prestige and style of the article. Discuss the impact of the article on the career today.
6. What hospital employees make up a hospital safety committee? Why should BMETs be on this committee? What role do they play in hospital safety and emergency planning?
7. How can BMETs promote good communication with staff about devices that do not work? Design and propose communication methods, including feedback from BMETs, that would enhance the relationship between BMETs and the staff who use the equipment.
8. Visit the Web site for the journal *24 x 7*. Look in the archives for an article that interests you. Summarize the article. Include a reflection on how this information might have an impact on your career when you are working in the field.
9. Explore generic Web sites that post employment ads, such as monster.com. Search for positions that are related to BMET work. Summarize the number and type of positions you find.

Now search Web sites that post positions for BMETs such as AAMI's job postings at http://www.aami.org/CareerCenter/SearchJobPostings.cfm. Describe how the opportunities are different. Visit major BMET employers such as http://www.philips.com (search in "Healthcare") and http://www.aramarkhealthcare.com and explore the employment sections. Summarize a position that you find that seems appealing to you.

2

Patient safety

LEARNING OBJECTIVES

1 define the types of currents related to the human body
2 identify the amount of current related to physical sensation, pain, injury, and death
3 define microshock and macroshock
4 define the hazardous currents in clinical electrical equipment
5 identify the basic AAMI recommended limits for currents in permanently wired devices and portable ones
6 identify NFPA 99 code
7 identify electrical receptacle requirements in a hospital (wiring and testing)
8 define GFCI and LIM and identify the regulations set for their performance in the clinical environment
9 define the patient care area
10 identify the maximum duration of power interruption before emergency power is provided
11 know NFPA 99 code requirements for extension cords and outlet strips
12 identify the code requirement for the ground-to-chassis resistance measurement

13 understand how NFPA 99 can be used to obtain maintenance manuals for equipment

14 identify the applicability of Life Safety Code 101

Introduction

The most important responsibility of BMETs relates to patient safety. Ensuring the safe use of technology is a vital role of the BMET as part of the medical care team. Understanding the human body and its reaction to externally applied voltage and/or current is vital to patient safety.

Electrical shock

An injury related to electrical shock may occur in any environment, but there is a higher potential for electrical injury in the hospital because of the direct contact of patient or caregiver and equipment. In addition, there are a great many devices that may be associated with one patient. The sensations or characteristic symptoms of various levels of electrical current are described in the following paragraphs and summarized in Table 2.1. The effects of electrical currents on the human body and tissue may range from a tingling sensation to tissue burns and heart fibrillation leading to death.

Electrical energy has three general effects on the body:

1. Resistive heating of tissue
2. Electrical stimulation of the tissue (nerve and muscle)
3. Electrochemical burns (for direct current)

TABLE 2.1. *Human detection of current*

Current description	Current (mA)	Physiological effect
Threshold	1–5	Tingling sensation
Pain	5–8	Intense or painful sensation
Let go	8–20	Threshold of involuntary muscle contraction
Paralysis	>20	Respiratory paralysis and pain
Fibrillation	80–1,000	Ventricular and heart fibrillation
Defibrillation	1,000–10,000	Sustained myocardial contraction and possible tissue burns

Note: These values will vary based on the person's gender, size and weight, skin moisture content, and pain tolerance levels.

When the human body is exposed to current, the reactions can be grouped based on the quantity of current. There are six current categories:

Threshold Current (1–5 milliamperes (mA)): This is the level of current required to perceive the feeling of current. A slight fuzzy feeling or tingling sensation is common at this current strength.

Pain Current (5–8 mA): This current level will produce a pain response, which may feel like a sharp bite.

Let Go Current (8–20 mA): This current level results in involuntary muscle contraction. Nerves and muscles are strongly stimulated, resulting in pain and fatigue. At the low end of let go current is the maximum amount of current from which a person can move away voluntarily (about 9.5 mA). At these levels, injuries may result from the instinct to pull away, for example, arm dislocation or broken bones from falls.

Paralysis Current (>20 mA): At levels greater than about 20 mA, the muscles lose their ability to relax. This includes the muscles involved in breathing. The breathing pattern can no longer be maintained and results in respiratory paralysis. Respiratory paralysis can result in death.

Fibrillation (80–1,000 mA): At levels between 80 and 1,000 mA, the heart goes into fibrillation. Fibrillation is the unsynchronized contraction of the muscle cells within the heart. During fibrillation, the heart is ineffective in pumping blood to the body. Heart fibrillation will result in death.

Defibrillation (1,000–10,000 mA): The delivery of electrical energy to the fibrillating heart is called defibrillation. A large current delivered by paddles at the skin, through muscle and bone, can resynchronize all of the cardiac muscles. Then, coordinated electrical generation can return to the heart. During open-heart surgery, spoon-shaped paddles can deliver much lower currents directly to the heart to induce fibrillation (to perform bypass surgery, for example) and then defibrillate the heart after the procedure is complete.

Researchers at Massachusetts Institute of Technology would like these current ranges redefined. One reason they would like to see the change is the vast differences among people in size and perception of pain. Because there is such a great variance from person to person, keep in mind that the current ranges listed here are only guidelines and approximations.

The electrical shock situations described in Table 2.1 are identified by the term macroshock. **Macroshock** is a physiologic response resulting from electrical current in contact with the skin of the body. Macroshock can occur when a person makes an electrical connection with two parts of the body (arms, for example) or a person is connected to "earth ground" (a lower potential area) and makes an electrical connection with one point of an energized source.

The skin of the body provides some protection from electrical hazards because of the skin's resistive properties. Dry, unbroken skin acts as an insulator. In the hospital environment, a patient may be especially susceptible to small electrical currents because the skin may be wet (patient fluids) from wounds.

One patient may have a fluid-filled catheter, and another may have an electrical wire in direct contact with her heart. If this pathway is used to conduct electricity directly to the heart, the patient can experience microshock. **Microshock** is a physiological response resulting from electrical current applied to the heart. Microshock currents are often tiny, so they are measured in microamperes (μA). Because there is a direct connection to the heart, even these small currents can be large enough to cause fibrillation of the heart. Be aware that very little can be done within a power system to protect against microshock. Isolated power, ground fault circuit interrupters, line isolation monitors, and other safety precautions do not protect the heart directly from these *low* current levels.

Leakage currents

All electronic devices have naturally occurring unintended currents within them. These are not due to any faults in the devices;

they are simply present. *All* devices have leakage current. For medical devices, BMETs categorize and measure these leakage currents.

Four categories of leakage currents are measured and have recommended safe limits. The categories are determined by the method through which a person might come in contact with the current or the device. The types of currents are:

- **Earth leakage current** (also called *earth risk current*) is the current that flows from the power supply of the device, across the insulation of the device, and through the ground in the power cord (three-prong power cord).
- **Enclosure leakage current** (also called *enclosure risk current* and *touch/chassis current, chassis leakage current*) is the current that would flow through a person if he touched any part of a device. The person could then form a connection to earth ground so this current is measured between the chassis and the earth ground.
- **Patient leakage current** (*patient risk current*) is the current that flows between the parts of a device that are in normal contact with the patient such as patient leads (on the skin or under the skin) and the earth ground.
- **Patient auxiliary current** is the current that can flow between two separate patient circuits or connections, like two different electrodes that both connect to the patient.

But what is leakage current and where does it come from? **Leakage current** is best defined as the small current that flows from the components of a device to the metal chassis. This is natural and is a result of wiring and components. It can be either resistive or capacitive. **Resistive leakage current** comes from the resistance of the insulation surrounding power wires and transformer windings. Resistive leakage current is much smaller than capacitive leakage current. **Capacitive leakage current** forms between two oppositely charged surfaces, such as between a wire and a chassis case or between two wires, one hot, one neutral. A capacitive current is formed between the two surfaces and tends to stray from the intended current path. Adding a safety ground wire is a method to reduce excessive leakage current. The third wire acts to divert the stray or leakage current away from the chassis (which the patient or caregiver may come in contact with) and to the intended circuit ground in the case of a short between a hot wire and a chassis ground.

Device currents are identified in two circumstances: when the device is working properly and when there is a fault. The **fault current** is the current that flows when the device is broken, the worst-case-scenario condition. The fault current is the maximum possible current flow from a device to ground or a person or another metal object. Types of faults occur when

- The ground is not connected.
- Each barrier of a double-insulated instrument is short-circuited.
- A supply conductor (hot or neutral) is not connected.
- Hot and neutral are reversed.
- A single component that can produce a hazardous current fails.
- Line voltage is applied to an input or output part or chassis (for ungrounded equipment).

- Line voltage is applied to a patient part (for isolated patient connections).

Standards

The Association for the Advancement of Medical Instrumentation and the National Fire Protection Association establish recommendations for the limits for leakage currents for medical devices. There are different limits for the two types of equipment connections. Current limits are higher for **fixed equipment** or **hard-wired equipment**, devices that are permanently wired to the electrical power supply (for example, imaging devices), than they are for devices that are "plugged in" using a power cord. Even though we use the term "fixed equipment," it should not be confused with equipment that has been repaired.

Summary of important values

For *fixed equipment*, earth leakage current is tested prior to installation when the equipment is temporarily insulated from ground. The leakage current from the device frame to ground of permanently wired equipment installed in general or critical patient care areas cannot exceed 5.0 mA (with grounds disconnected). For portable equipment, earth leakage current cannot exceed 300 μA (raised from 100 μA in 1993).

When *multiple devices* are connected and more than one power cord supplies power (for example, equipment located on a cart), the devices should be separated into groups according to their power supply cord, and the leakage current should be measured independently for each group as an assembly.

NFPA 99

The National Fire Protection Association established a code in 1984 "to minimize the hazards of fire, explosion, and electricity in health care facilities providing services to human beings." Code 99 applies to health care and the technology used within the clinical setting. The entire code is available online at http://www.nfpa.org. Some states have adopted NFPA 99 as state law. When states adopt these standards as law, they become absolute legal requirements. Currently, the American Society for Healthcare Engineering is working to rewrite this code. More information is available at http://www.ashe.org, click on "Codes and Standards."

Standard regulatory definitions

Anesthetizing locations: An area of the facility designated to be used for the administration of nonflammable inhalation anesthetic agents. (Flammable anesthetics, such as ether, are no longer used in this country. There were many precautions and regulations when flammable anesthetics were used.)

Fault current: A current due to an accidental connection between the energized (hot) line and ground.

Grounding system: A system of conductors that provides a low-impedance return path for leakage and fault currents.

GFCI (ground fault circuit interrupter): A device that will *interrupt* (this means "break" or "stop power") the electric circuit to a load when the fault current to ground exceeds a specified value. GFCIs must activate when fault current to ground is 6 mA or greater.

Isolated power: A system that uses a transformer to produce isolated power and a line isolation monitor. The electrical grounds of the system are connected (on both the hospital and isolated side). It is the power that is isolated (hot and neutral).

Isolation transformer: A transformer used to electrically isolate two power systems.

LIM (line isolation monitor): A line isolation monitor is an in-line, isolated-power test instrument designed to continually check the impedance from each line of an isolated circuit to ground. It contains a built-in test circuit to test the alarm without adding additional leakage current. The line isolation monitor provides a warning (usually a loud buzz or other noise) when a single fault occurs or when excessively low impedance to ground develops, which might expose the patient to an unsafe condition should an additional fault occur. When the total hazard current reaches a 5-mA threshold, the monitor should alarm. LIMs must alarm when the fault current (from conductor to ground) is 5.0 mA or greater. LIMs must *not* alarm if current is 3.7 mA or less. Be aware that the LIM does *not* break the circuit like a GFCI. Excessive ground currents trigger alarms but will not stop the power to the system.

LIMs must be tested after installation and *every 6 months* by grounding hot and neutral through a resistor. Also, there must be a check of the visual and audible alarms. The LIM must be tested

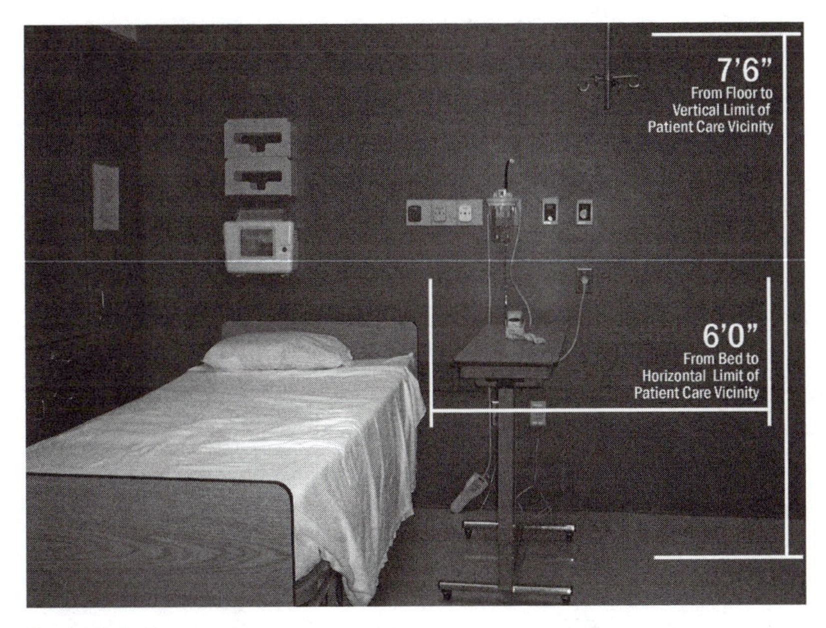

Figure 2.1. Patient care area.

each month with the test button. LIMs with automated self-test capabilities must be tested *every 12 months.*

Medical air: Air that is 19.5–23.5% oxygen. It also has specifications that limit contaminants such as moisture or bacteria.

Patient care location: Any portion of a facility where patients are intended to be examined or treated. This environment is defined around the patient bed. The **vicinity** is defined as 6 feet around the bed and 7 feet 6 inches above the floor below the patient. See Figure 2.1.

Patient connection: An intended connection between a device and a patient that can carry current. This can be a conductive surface (for example, an ECG electrode), an invasive connection (for example, an implanted wire), or an incidental long-term connection (for example, conductive tubing). This is *not* intended to include casual contacts such as push buttons, bed surfaces, lamps, and hand-held appliances.

Wet location: A patient care area that is normally subject to wet conditions while patients are present (not routine housekeeping). An example would include physical therapy areas where whirlpool baths are used.

Electrical wiring in the hospital

Interestingly, outlets in a hospital are usually wired with the ground pin at the "top" of the outlet (see Figure 2.2). This is an NFPA 99 requirement. In states where NFPA is law (it is law in only about half of the United States), outlets are wired in this orientation. However, older facilities and hospitals in states where NFPA is not law may have outlets wired with the ground pin below the hot and neutral pins. Outlets need to be hospital grade. These outlets, which have a green dot stamped on the face, undergo rigorous testing. Prior to 1996, electrical outlets in hospitals had to be checked (integrity, polarity, and ground connection) once per year. In 1996, the regulations regarding the frequency of testing of receptacles was changed. No time interval is now specified *except* for non-hospital-grade receptacles, which must be tested at least once every 12 months. Therefore, most hospitals use hospital-grade receptacles (in patient care areas) so that they do not need to be checked. Electrical outlets also have a specified amount of "holding power" for the ground pin. The code requires retention force of grounding blade of not less than *4 ounces*.

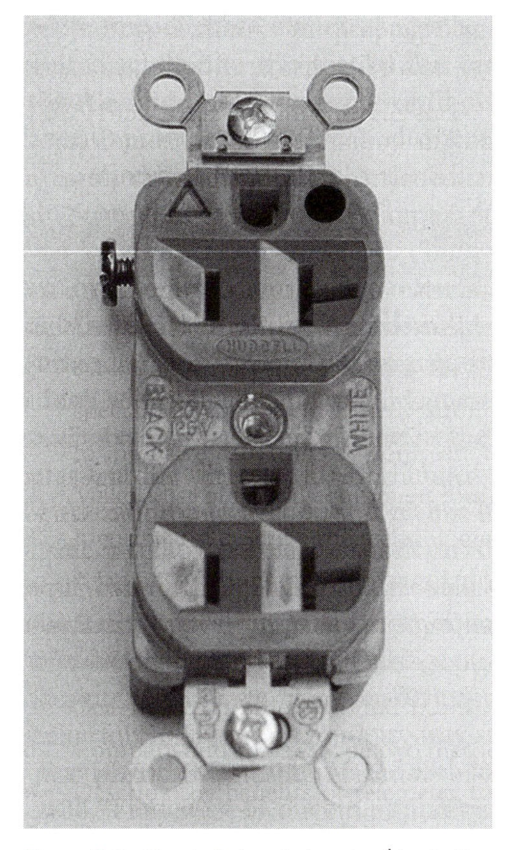

Figure 2.2. Hospital electrical outlet (dot indicates hospital grade).

Some hospital outlets are colored red (either the receptacle itself or its cover faceplate). This indicates that the outlet is wired to emergency power (see emergency power time specifications) in the event of an outage. Typically, life-saving devices, such as ventilators, are connected to the emergency red outlets.

Wire gauge for power cord ground conductor: If the length of the power cord is less than 15 feet, use 18 AWG; if the length is greater than 15 feet, use 16 AWG.

Standard colors for wiring: The black = hot, white = neutral, and green = ground.

Placement in the clinical laboratory: Electrical outlets must be installed every 1.6–3.3 feet.

Ground impedance: Both capacitive reactance and resistance between either the hot or the neutral conductors and ground must exceed 200 kilo ohms (kΩ) in an isolated system. This must be tested in 10% of outlets during new construction.

Emergency power: Power must be available within *10 seconds.* Electrical outlets connected to emergency power are colored red (or their faceplates are labeled "emergency power").

Extension cords are allowed to be used within the hospital; however, three-to-two-prong adapters are not allowed.

Two-wire (ungrounded) power cords: The NFPA code states that all devices must have three-wire power cords with a three-prong plug. However, devices that are double insulated can have a two-conductor cord and two-prong plug. A device that is double insulated will be marked on the case using a square inside a square symbol.

Extension cords in operating rooms: According to the code, the power cord from a device must be "continuous and without switches from appliance to the attachment plug" (NFPA 99 7-5.1.2.5). This has been interpreted as *forbidding* extension cords in anesthetizing locations (which is generally the same as operating rooms). The only exception is a permanently mounted power cord on a movable equipment assembly like a rack or table (an outlet strip).

(An **extension cord** is defined as a cord that has a male connector on one end and a female connector on the other end and that is intended to allow the *power cord* from a device to be connected to a power outlet further in distance from the device than the length of the device power cord.)

Please note that there is no limit to the length of a power cord for a device in operating rooms. Some hospitals replace shorter power cords with longer ones to eliminate the "too far from the outlet" issue. This also diminishes the need for extension cords.

Outlet strips in operating rooms: The code states that it is possible to use a hospital-grade outlet strip as long as it is permanently mounted to a rack or table. In addition, the total current drawn by all of the appliances cannot exceed 75% of the current capacity of the strip.

Ceiling-mounted receptacles and drop cords in operating rooms: The receptacles are the twist-lock type.

Ground-to-chassis resistance measurement: The resistance between the device chassis, or any exposed conductive surface of the device, and the ground pin of the attachment plug shall be measured. The resistance shall be *less than 0.50 ohm (Ω).*

Manuals

NFPA specifically requires that the vendor supply suitable manuals for operators and users upon delivery of the appliance. "Purchase specifications shall require the vendor to supply suitable manuals for operators and users upon delivery of the appliance. The manuals shall include installation and operating instructions, inspection and testing procedures and maintenance details" (7-6.2.1.8). This is often cited in a purchase agreement as a method for the hospital to obtain maintenance materials and circuit diagrams from the manufacturer.

The manuals must include:

- Illustrations that show location of the controls
- Explanation of the function of each control
- Illustrations of proper connection to the patient and other equipment
- Step-by-step procedures for proper use of the appliance
- Safety considerations in application and in servicing
- Difficulties that might be encountered, and care to be taken if the appliance is used on a patient simultaneously with other electrical appliances
- Schematics, wiring diagrams, mechanical layouts, parts lists, and other pertinent data for the appliance as shipped
- Functional description of the circuit
- Electrical supply requirements, weight, dimensions, and so on
- Limits of supply variations
- Technical performance specs
- Instructions for unpacking
- Comprehensive preventative and corrective maintenance and repair procedures

The NFPA Life Safety Code 101

The NPFA Line Safety Code 101 is meant to provide minimum regulations to protect occupants of many types of buildings including day care facilities, high-rise structures, stores, factories, and hospitals. The code covers:

- Means of egress
- Occupant notification (fire alarms and drills)
- Interior finish, contents, and furnishings
- Fire protection equipment
- Building services

The code contains a great deal of information necessary for safe design and construction of the facility. The code is different for new and existing heath care facilities. This information can be summarized as follows:

- Compartmentation of fire is required because evacuation cannot be ensured.
- Automated sprinklers are required for new construction and major renovations.
- The number of floors permitted is related to the type of construction (for example, the exterior of a building with four or more floors must be made from a noncombustible material or limited combustible material).
- Planning for the relocation of patients in a fire must include movement of patients in their beds.
- Doors to patient rooms are not permitted to lock, except where the needs of the patients require locked doors and keys are carried by the staff at all times.
- Doors in most corridors and walkways may only be held open by automatic release devices that close in case of fire; in order to avoid staff using objects to hold doors open, doors should be equipped with automatic hold open devices.
- Two exits must be provided from each floor.

- Most service rooms (housekeeping, facilities) must have a one-hour fire rating.
- Every patient room must have a window.

Government regulations

The **Medical Device Amendment,** May 28, 1976, gave the Food and Drug Administration (FDA) authority to regulate medical devices. The **Safe Medical Device Act of 1990 and Amendments of 1992** require hospitals to identify and report serious problems with medical devices.

Hospitals must file an MDR (which stands for medical device reporting) when it is determined that a device has caused or contributed to a patient death or serious injury. These reports must be submitted within 10 days from the time personnel become aware that the device has caused or contributed to injury or death. An online database of these reports is available and can be searched.

What does health insurance reform have to do with medical devices? The **Health Insurance Portability and Accountability Act of 1996 (HIPAA)** has requirements regarding privacy of medical information. Since patient information is often stored in the memory of medical devices, some BMETs work to ensure hospital compliance with the privacy of the information. They make sure there are limitations regarding who has access to the stored data as they carefully work to protect patient privacy.

STUDY QUESTIONS

1. Describe a scenario that would result in macroshock. Identify the power source.

2. Describe a scenario that could result in microshock. Identify the power source and describe how the current travels through the patient.
3. Describe the conditions under which electrical current can be used to correct the electrical signals of the heart.
4. Make a list of some of the guidelines related to equipment power cords.
5. Make a list of some of the guidelines related to power outlets.
6. Make a chart that shows the currents necessary to trigger the action of a LIM and a GFCI. Include a column that shows whether the circuit is open or closed above the trigger currents.
7. Summarize NFPA 101 requirements.
8. Describe the role of the federal government in medical equipment regulations.
9. Summarize how HIPPA impacts what BMETs do.

FOR FURTHER EXPLORATION

1. Leviton is a manufacturer of hospital-grade outlets and plugs. Use its brochure *Industrial Wiring Devices for the Health Care Industry* (available on the Internet; use quotes around the title in a search engine to locate the correct page) to summarize the testing required in order to be labeled "hospital grade." Describe why this is useful for outlets designed for the hospital environment.
2. There has been some discussion about UL regulations for relocatable power taps (outlet strips) and the rules that apply to hospitals. The confusion has prompted a new specific exclusion for hospitals. Summarize this ruling, which can be found at http://ulstandardsinfonet.ul.com/scopes/1363.html (scroll down to the end of the page).

3. What is the major difference between a LIM and a GFCI? Why are LIMs (instead of GFCIs) specifically used in the operating room?
4. Why do you think the regulation for power strips limits the items connected to the strip to 75% of the strip's rated value? Describe how this could be implemented/restricted in the clinical setting.
5. Twist lock power cords are required for power cords that drop from the ceiling. This is fairly common in the operating room to limit power cords on the floor. Use the Internet to research the look of twist lock connectors. Describe the benefits they offer to the situation over common prong electrical connections.
6. A common set of questions on certification exams includes the numerical data points set by NFPA. Prepare a list of such requirements including:
   - ▷ Leakage current for portable equipment
   - ▷ When a GFCI must trip
   - ▷ Retention force of grounding blade
   - ▷ When an LIM must alarm
   - ▷ Wire gauge requirement for power cord ground conductor
   - ▷ Availability of emergency power
   - ▷ Ground-to-chassis resistance measurement
7. Use the Internet to research the history of the FDA and its involvement in medical device design, manufacture, and use. Summarize the history including historical milestones. Provide your opinion as to whether the FDA has been able to "keep up" with regulations in light of changing technology.
8. Explore the database of device problems on the FDA site: The address for a search is http://www.accessdata.fda.gov/scripts/cdrh/cfdocs/cfmdr/search.CFM. To enter a search string, use a device name from this text. Search for an incident and summarize it. Discuss the role of human error and device flaws (in design or function) in this incident.

3

In the workplace

LEARNING OBJECTIVES

1. list, describe, and characterize applicable codes of ethics, with patient safety and confidentiality as primary concerns
2. list and describe commonly used test equipment
3. list and describe safety universal precautions and personal safety measures
4. list and describe the qualities of excellent customer service

Introduction

As part of the health care team, BMETs must strive to serve the staff and patients in the safest and most effective ways possible. This chapter reviews the various ethical codes that may apply to BMETs. In the health care workplace, test equipment is used to ensure patient safety. Precautions are employed to promote personal safety. In addition, serving staff and patients requires excellent customer service skills. These important facets of the BMET workplace are explored.

Applicable codes of ethics

No code of ethics specifically and exactly addresses the needs of hospital-based BMETs. Two codes, however, have parts that apply to BMETs. First, there is the Code of Ethics from the **Biomedical Engineering Society** (http://www.bmes.org). Even though this group is research oriented, it does have a section that applies to BMETs in the hospital environment.

> **Biomedical Engineering Professional Obligations**
> Biomedical engineers in the fulfillment of their professional engineering duties shall:
> 1. Use their knowledge, skills, and abilities to enhance the safety, health, and welfare of the public.
> 2. Strive by action, example, and influence to increase the competence, prestige, and honor of the biomedical engineering profession.

Biomedical Engineering Health Care Obligations

Biomedical engineers involved in health care activities shall:

1. Regard responsibility toward and rights of patients, including those of confidentiality and privacy, as their primary concern.
2. Consider the larger consequences of their work in regard to cost, availability, and delivery of health care.

A second group has developed an applicable code of ethics. The **American College of Clinical Engineering (ACCE)** has established the following guidelines:

Preamble

The following principles are established to aid individuals practicing engineering in health care to determine the propriety of their actions in relation to patients, health care personnel, students, clients, and employers. In their professional practices they must incorporate, maintain and promote the highest levels of ethical conduct.

Guidelines principles of ethics

In the fulfillment of their duties, Clinical Engineers will

- Hold paramount the safety, health, and welfare of the public
- Improve the efficacy and safety of health care through the application of technology
- Support the efficacy and safety of health care through the acquisition and exchange of information and experience with other engineers and managers
- Manage health care technology programs effectively and resources responsibly

- Accurately represent their level of responsibility, authority, experience, knowledge and education and perform services only in their area of competence
- Maintain confidentiality of patient information as well as proprietary employer or client information, unless doing so would endanger public safety or violate any legal obligations
- Not engage in any activities that are conflicts of interest or that provide the appearance of conflicts of interest and that can adversely affect their performance or impair their professional judgment
- Conduct themselves honorably and legally in all their activities

This can be found on the Web site at http://www.accenet.org/; at the "Publications and Practices" tab, select "Professional Practices."

To summarize these codes, the essential foundation and guiding principle puts patient safety as primary in all activities. In addition, patient privacy is paramount. The Health Insurance Portability and Accountability Act of 1996 compels BMETs to ensure patient privacy.

Test equipment for the BMET

Many different types of tests must be performed on medical equipment. To accomplish this, there are devices specifically designed to check the performance or safety of this equipment. Some test equipment can simulate patient physiological signals and are used to verify proper performance of devices.

Some companies produce equipment for most testing needs of BMETs. Major manufacturers include BCGroup, Dale Technology, Metron, and Fluke Biomedical. Some manufacturers provide very specific testing devices made to interface with specific pieces of equipment.

Guidelines of best practice and some regulations may require testing of equipment for chassis leakage and ground resistance. Also, outlets must be checked for polarity under certain conditions. Devices that will automatically check and then measure these variables in the specified conditions are often called **safety analyzers** (see Figure 3.1).

Some specialized test equipment is designed to perform various kinds of recommended tests. For example, patient leads for cardiac monitors must be tested for electrical isolation. Numerous test devices that will perform this test are available. BMETs can also use a specialized device to test GFCIs. Defibrillators and electrosurgical units must also be tested for proper output. There are specific devices designed to test these pieces of equipment. Some of the types of test equipment are made by the manufacturer just for a particular machine; others are more generic. Devices like IV pumps, ventilators, pulse-ox devices, and ultrasound machines all require performance testing, and many have unique devices to check them. Figure 3.2 shows a BMET testing the output of a defibrillator that is part of a "crash cart." During cardiac emergencies, crash carts contain items that can be used to resuscitate a patient. Proper performance of defibrillators in emergencies is essential and so they are regularly tested.

BMET personal safety

Universal precautions and isolation

You must assume every patient and every piece of equipment has a communicable disease that could be transmitted to you.

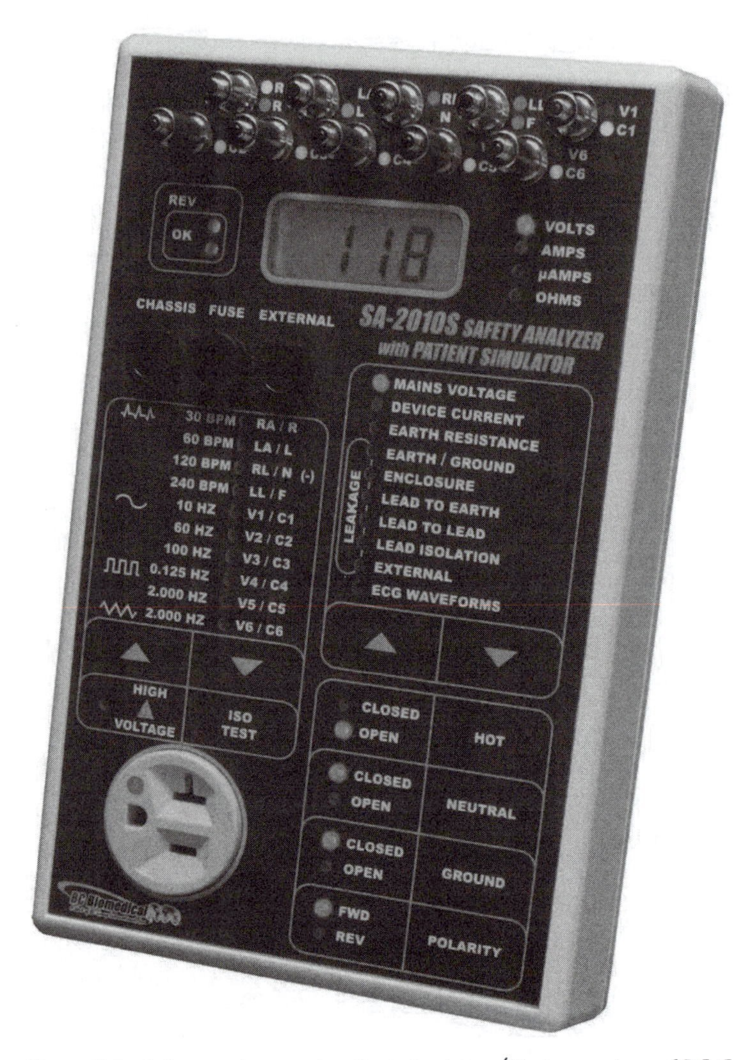

Figure 3.1. Safety analyzer and patient simulator. (Photo courtesy of BC Group.)

Using **universal precautions** treats all human blood and certain human body fluids as if they are infectious for HIV and other pathogens. In general, AIDS is the only infectious disease for which a patient cannot be tested without patient permission.

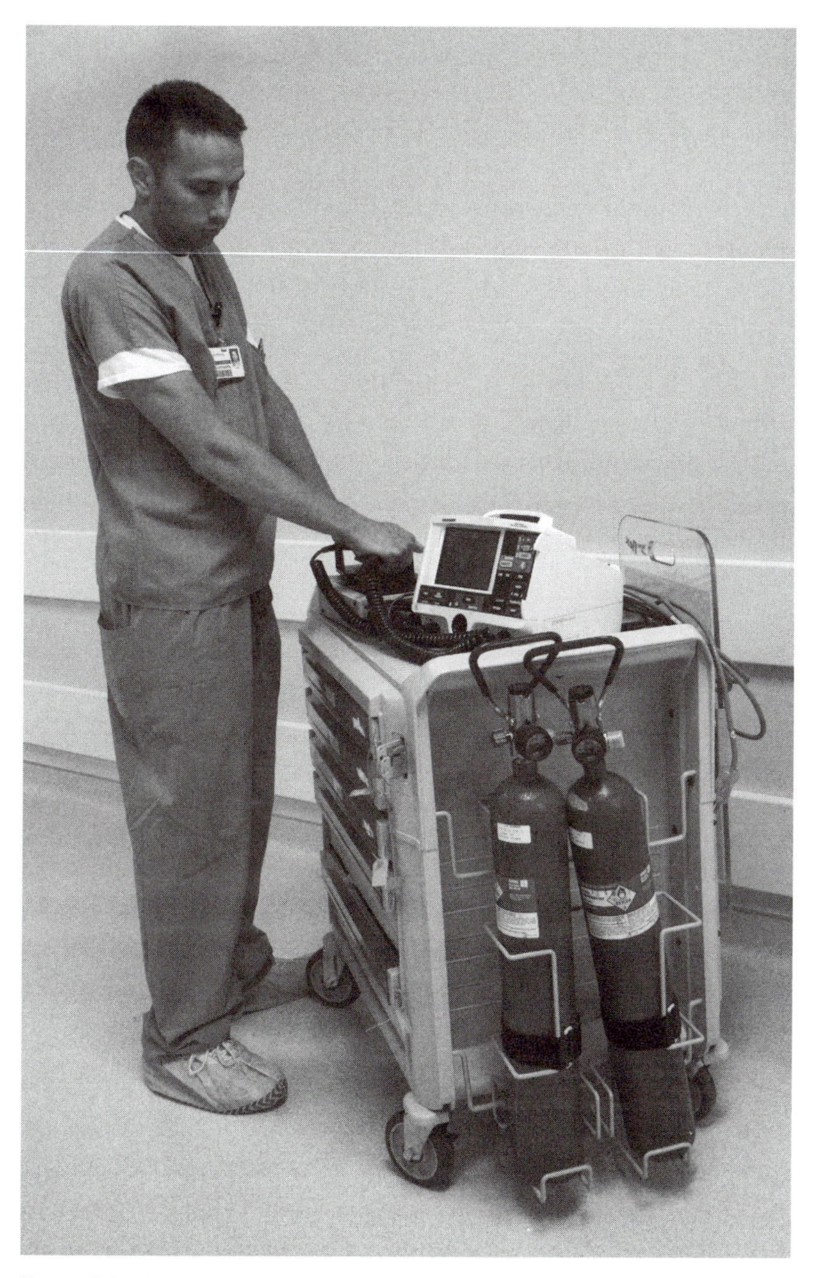

Figure 3.2. A BMET testing a defibrillator.

Consequently, hospitals assume all patients and all fluids associated with the patient are contaminated. Almost all bodily fluids and some body parts are systematically treated with the utmost care. The bodily fluids on the precautions list include semen, vaginal secretions, cerebrospinal fluid, synovial fluid, pleural fluid, pericardial fluid, peritoneal fluid, amniotic fluid, saliva in dental procedures, any body fluid that is visibly contaminated with blood, and all body fluids in situations where it is difficult or impossible to differentiate between body fluids. Interestingly, sweat is not included in this list. The body parts on the precautions list include the mucous membranes and any body tissue or organ (other than intact skin) from a human (living or dead).

Precaution standards

The Centers for Disease Control and Prevention (CDC) has identified two basic classes of standards: standard precautions and transmission-based precautions.

Standard precautions are to be considered and used for every patient and every piece of patient equipment encountered. Standard precautions combine Universal precautions (blood and body fluid precautions) and the use of personal protective equipment (**PPE**) to provide a barrier between moist body substances (for example, blood, sputum, and secretions) and health care workers and visitors. PPE may include gloves, glasses, gowns, face shields (see Figure 3.3), and masks. Use of standard precautions takes into consideration that sources of infection may be recognized or unrecognized as infectious at the time of exposure.

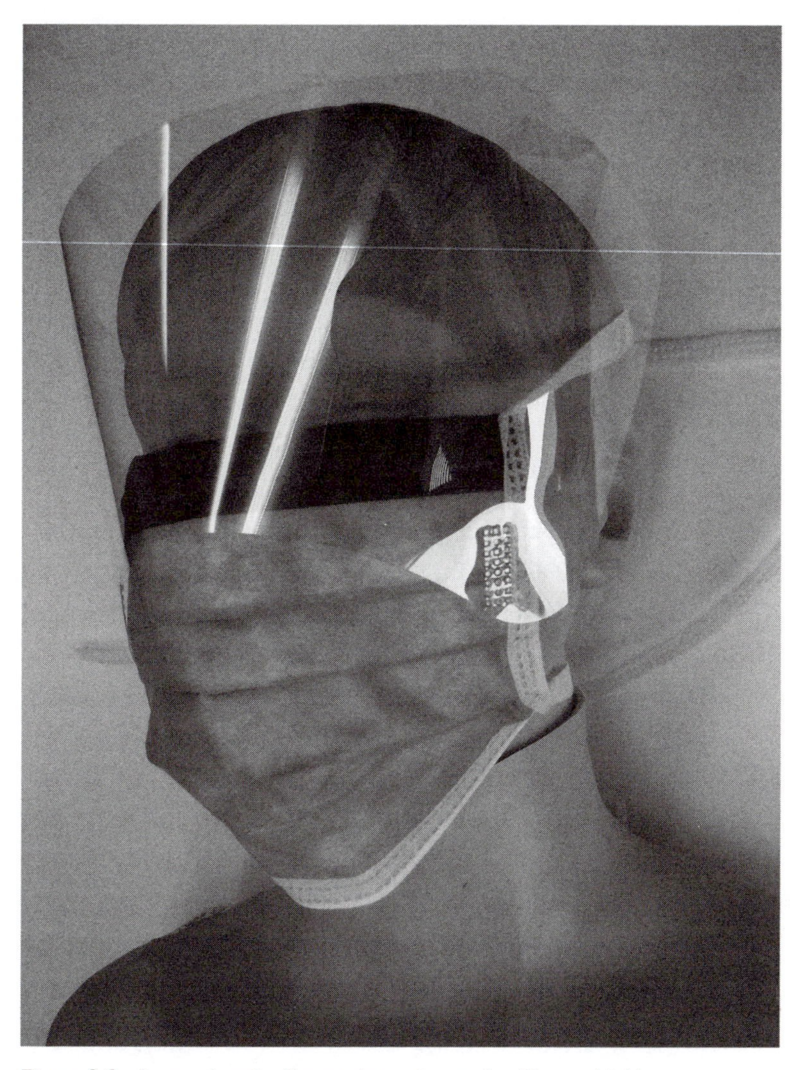

Figure 3.3. Personal protective equipment – mask with eye shield.

Transmission-based precautions are designed for patients documented or suspected to be infected or colonized with highly transmissible pathogens for which additional precautions are needed to interrupt transmission. These precautions are used in addition to standard precautions. Transmission-based precautions include airborne, droplet, and contact isolation. You need to learn the facility-specific signage associated with each of these isolation types so that you are aware of the practices required before entering a patient care environment.

- **Airborne precautions** - Follow these precautions around patients known or suspected to have serious illnesses transmitted by airborne droplet nuclei. Examples of such illnesses include measles, varicella (chicken pox), and tuberculosis.
- **Droplet precautions** - In addition to standard precautions, use droplet precautions for patients known or suspected to have serious illnesses transmitted by large particle droplets. Examples of such illnesses include invasive *Haemophilus influenzae* type b disease, including meningitis, pneumonia, epiglottitis, and sepsis, and invasive *Neisseria meningitidis* disease.
- **Contact precautions** - In addition to standard precautions, use contact precautions for patients known or suspected to have serious illnesses easily transmitted by direct patient contact or by contact with items in the patient's environment. Examples of such illnesses include gastrointestinal, respiratory, skin, or wound infections or colonization with multidrug-resistance.

Customer service

When a BMET has a positive experience with a clinical staff member, especially in a time of difficulty, it can greatly improve future interactions and positively impact patient care. Understanding how to approach stressful situations and communicate effectively is a learned skill that can be perfected over time. However, some important skills, specific to BMETs, can be identified.

- Exercise generally accepted best practices of effective customer service. These include compassion, friendliness, advocacy, commitment to excellence, flexibility, and willingness to help.
- Be responsive, respond to phone calls, pages, and email messages. Respond as quickly as possible, even if it is only to arrange follow-up at a later time.
- Initiate ongoing communication. Visiting units at times when there is not a problem can build rapport and trust between the BMETs and staff.
- Listen to a staff member's interpretation of a problem, even when it's improbable; it can build communication for future scenarios – when the staff member might be correct.
- Understand that the stress on the medical staff due to a technological malfunction may be enormous because their ability to treat the patient is compromised. What may seem minor to a BMET may be huge to the medical staff.
- Be sensitive about communicating to the device user the source of a problem after a specific solution is identified, especially if the staff may have overlooked a simple solution in the stress of the moment. It may be a good idea to wait until stressful emotions have passed to communicate the overlooked step. Using the opportunity to explain how simple situations escalate can avoid repetition of the problem.
- Identify processes that can avoid the simplistic communication of "broken." One word descriptors are not very useful in BMET problem solving.

- Be constantly aware of opportunities to clarify equipment function to avoid problems. If a problem seems to occur frequently, staff instruction or procedural changes can decrease the difficulties.
- Do your homework! Keep up with the treatment trends and technologies. Make use of all available resources including the Internet to be knowledgeable and helpful. Technology is always evolving, and BMETs need to know where to obtain information when presented with a device or situation with which they are unfamiliar.

One of the most valuable skills you can acquire is an ability to look at the human body without giggles or disgust. Better said, BMETs need to have a **professional attitude** and calm composure when working around patients. Prior to a hospital career, most people have little exposure to the naked adult human body. The clinical environment requires an entirely new perspective. Effective and professional BMETs are calm and focused to the best of their abilities, often in stressful situations. The human body is not "pretty" in trauma and disease. The composure of a BMET, as a valued member of the health care team, is the most important item in the BMET tool kit.

STUDY QUESTIONS

1. Summarize the primary ethical qualities for BMETs that are reflected in the various codes of ethics.
2. Identify and describe some of the equipment tests that are performed by BMETs.

3. Describe a scenario in which a BMET would use PPE and be aware of the potential of patient fluid contamination.
4. Define and describe standard precautions.
5. Describe the benefits of effective customer service interactions with medical staff.
6. Identify some of the methods BMETs use to keep up to date with equipment and treatment innovations.
7. List and describe the qualities of a BMET that illustrate the professional attitude recommended in the chapter.

FOR FURTHER EXPLORATION

1. What happens to a BMET if he or she does not act ethically? Is there a consequence for violation of the HIPPA right to patient privacy? Since there is no BMET "license" that can be removed for unethical behavior, what other means are there to ensure that BMETs act responsibly? Use the Internet to research your response and document your answer.
2. Discuss the advantages and disadvantages of a license requirement for BMETs to practice in the clinical setting. Identify medical staff who are required to obtain a license (excluding doctors and nurses). Are there other medical staff *who do not work directly with patients* who need a license? Use the Internet to research your response and document your answers.
3. Browse through the Web sites of the test equipment companies to view some of the specialty testing devices. Summarize the types/categories of equipment available. How would a BMET department select devices to purchase? Discuss the importance of high-quality functioning test equipment in device support.

4. Some manufacturers restrict the purchase of specialized test equipment until service training has been completed by a BMET from the hospital. Why would this be a requirement? What are the advantages and disadvantages of this restriction? Discuss the economic impact of this type of restriction.
5. If one BMET from a department is sent for specialized training on a device, discuss the difficulties that person might experience in supporting all the devices all the time. Describe some of the ways a department could share the knowledge to share the equipment support.
6. *Patient isolation* is one method of transmission restriction. Explore the CDC Web site for the standards for patient isolation: http://www.cdc.gov/ncidod/dhqp/gl_isolation.html. Summarize the information found there. Carefully identify the ways a BMET would know a patient was subject to isolation.
7. Explore general customer service skills information available on the Internet. Summarize some of the basic points. How does the general information, say for retail sales, apply to the clinical setting? In what ways are customer service skills more important in a clinical setting than a retail establishment? Identify how a BMET can obtain customer service training.

4

Electrodes, sensors, signals, and noise

LEARNING OBJECTIVES

1. list and describe what a sensor does
2. identify the two types of transducers and describe examples of each type
3. list and describe the sources of error in sensor systems
4. list and describe the four types of electrodes (surface, micro, indwelling, and needle)
5. list and describe the sensors that are used to record direct blood pressure and temperature
6. list and describe impedance matching and patient impedance
7. list and describe human periodic, static, and random signals
8. characterize human signals as analog and medical devices as digital
9. describe electrical noise, especially related to the clinical setting

Introduction

The most common activity in patient care is patient monitoring. The human body produces a variety of physiological signals, and physicians have learned to interpret these signals to provide information about the health of a patient. To measure or monitor these signals, there must be a connection between the patient and some (typically electronic) device. This type of activity is called *in vivo* monitoring, which refers to the living patient. This chapter explores the machine-human connection.

Sensors

Sensors (often called transducers – the difference is unimportant here) convert the energy of the patient (pressure, for example) into a form that can be used by an instrument. There are *two types* of sensors/transducers:

- Some transducers' output changes in response to a change in surroundings. The output is a change in **resistance, capacitance, or inductance**. These variations can then be measured, often using Wheatstone bridge circuitry because the changes may be very small. Common examples include:
  - **Strain gauges** – Change in resistance when some external event occurs.
  - **Potentiometers** (variable resistors) – Change in resistance when some external event occurs. Since these devices often have a mechanical knob or lever, they often convert mechanical movement or position into a change in resistance.

  - ▷ **Thermistors** – Change in resistance when temperature changes.
  - ▷ **Photoresistors** – Change in resistance when light hits the device.
- Some transducers **produce a voltage or current** in response to a change in environment. Common examples include:
  - ▷ **Piezoelectric crystals** – Produce a voltage as the crystal is compressed (even tiny amounts).
  - ▷ **LVDTs** (linear variable differential transformers) – Convert linear motion (may be very small amounts) into an electrical signal.
  - ▷ **Thermocouples** – Measure temperature differences using dissimilar metals. They require a known reference temperature.

Note: There are *many* types of sensors/transducers. Commonly seen ones are described here, but engineers create specialized and unique devices to serve particular needs.

When **blood pressure** is measured directly (in an artery), it is commonly measured using strain gauges, piezoelectric crystals, and silicon membranes. The hospital environment is very physically demanding on equipment. Even though transducers are usually used only once (disposable), a good transducer must be very durable.

Optical devices

Photoemissive devices

Photoemissive devices are common and useful because of the relationship between electrical behavior and light. They are able to create light under certain conditions.

A **light-emitting diode (LED)** is a semiconductor device that emits light when it is biased in the forward direction. The color of the emitted light depends on the chemical composition of the material used in the semiconductor. It can be visible light (see the list of colors in the next paragraph), infrared, or nearly ultraviolet. LEDs are used as indicators in many devices. A green indicator may indicate that power is applied, while a red LED could indicate a nonfunctioning circuit. Applications are manufacturer-specific.

LEDs emit the following different visible colors:

- Red
- Orange
- Yellow and "amber"
- Green
- Blue-Green
- Blue
- White

Photodetectors

These devices can respond to light by transforming light energy under certain conditions. There are two types of photodetectors: photocells and photodiodes.

- **Photocells** convert light into energy (usually a voltage or current). A light-sensitive cathode emits electrons when light hits it. The emitted electrons are then collected at the anode, which results in electrical current. Photocells are also used as detectors in spectrophotometry.
- **Photodiodes,** also called light-detecting diodes (LDD), are made of silicon that generates an electrical signal (usually a

voltage) when light hits the diode. Photodiodes are also used as detectors in spectrophotometry, pulse oximetry, optical glucose meters, and laboratory equipment that measures blood plasma elements.

Important sources of error in systems with sensors

There are many opportunities for systems to incur error. Here are some general and common types of error and opportunities for unreliability:

Insertion error: The sensors interfere with the system and affect either the measurement or the system. This is often the case when the item that links the equipment to the patient causes some distortion or change in the signal to be measured. (For example, when patients undergo sleep analysis, they must sleep with a large amount of sensors attached to them. This may make the sleep an inaccurate representation of the typical sleep patterns.)

Environmental error: The hospital is a very abusive place: It is "rough" on sensors (they are dropped, for example). Patients and devices are often moved from place to place. Also, many staff members use the equipment on many different patients. For example, perhaps the environmental temperature in one situation is very different from that in another situation.

Hysteresis: It is difficult to have a sensor behave exactly the same way when "loaded" as it does when "unloaded." For example, if you measure three temperatures – one low, one high, and then one low – will the sensor be able to return to the low temperature accurately the second time, or will the exposure to the high temperature limit the device's ability to take a low reading?

Electrodes

Many important human physiological signals are electrical. An electrical connection is necessary to connect the patient to the device. **Electrodes** are often used to pick up the signals.

Of all the connections made, **surface electrodes** are the most common. Surface electrodes make the connection to the patient, *in vivo*, which means on living people. These are used in electrocardiograms (heart; ECGs or EKGs), electromyograms (muscle; EMGs), and electroencephalogram (brain; EEGs). These electrodes often use a metal conductor surrounded by a conductive, jelly-like material (usually embedded in a sponge-like material) and some sticky plastic foam to hold these components in place. See Figure 4.1. These electrodes do not pierce the skin.

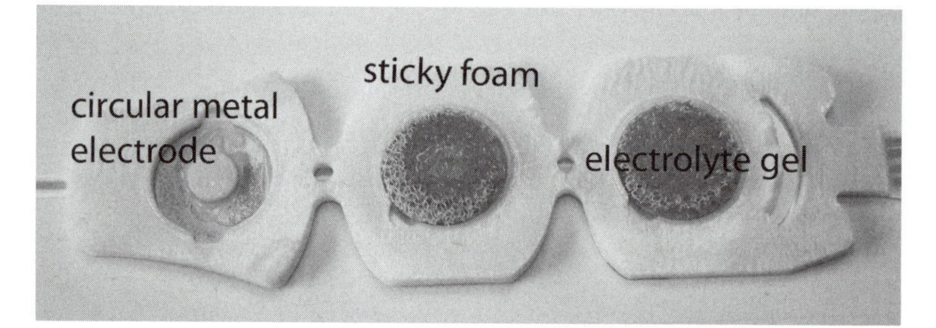

Figure 4.1. Surface electrodes (the electrolyte-soaked sponge has been removed from the electrode on the left).

The major difficulties with surface electrodes are:

- "Stick" problems - how to attach them to the patient
- Motion of the patient
- High impedance of human skin, especially if it is dry or hairy
- Patient environment - some patients may be in a wet environment, for example, which can hinder the electrode connection
- Available skin surface area - neonates and burn patients can present challenges because they have limited available skin for connection

These difficulties can diminish the signal strength or distort the signal in a way that makes it difficult to use for the diagnosis of patient conditions. As a result, efforts are made to diminish the impact of these situations. For example, hairy skin may be shaved to promote good electrode contact. Many devices have built-in algorithms that reduce the impact of motion on the signal. Research into the creation of "better" surface electrodes is ongoing.

Table 4.1 lists some types of electrodes. A good example of the use of a **needle electrode** is in fetal monitoring (see Figure 4.2). An electrode is attached to the head of a fetus before delivery. **Scalp electrodes** penetrate the skin of a baby prior to delivery. The long wire has a metal spiral tip that is twisted into the scalp in order to obtain ECG readings. For high-risk deliveries, this is an important monitoring technique.

Measuring temperature

Temperature is a very common variable to be measured. Patient temperature is very important in the assessment of patient

TABLE 4.1. *Other types of electrodes*

Types of electrodes	What signal is the electrode used to measure?	How is the electrode attached to the patient?
Needle electrode	EEG and ECG	Slightly below the skin
Indwelling electrodes	Specialized heart signals	Often threaded through the patient's veins
Microelectrode	Cell potentials	Not often attached to patients, often used in research

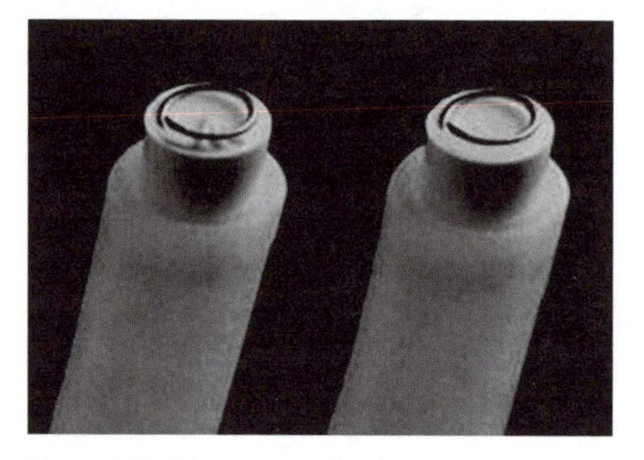

Figure 4.2. Microelectrodes that twist into the scalps of neonates during delivery.

condition. Typical patient temperature ranges are shown in Table 4.2. The core human body temperature can be influenced by many things including pregnancy, smoking, the environment, and other conditions that influence the body's ability to control its own temperature. *Thermistors* are often used to measure skin or internal patient temperature. They are inexpensive, relatively rugged, and nonlinear. Circuitry that processes the sensor's data can compensate for the nonlinearity. *Thermocouples* that require a reference temperature are also used, as are solid-state temperature sensors.

TABLE 4.2. *Typical patient temperatures*

Hypothermia	Normal body temperature (core)	Fever/hyperthermia
86–95°F	97.5–98.9°F	99–107°F

Impedance matching

To pass a signal from the patient to the device, the patient and the device input must have the same (or close) impedance. Since the *patient is usually a high-impedance* source, amplifiers are needed to match the high impedance. Devices are designed with this impedance matching in mind.

Human signals

The human body produces a wide variety of electrical (and other types of) signals, which are a perfect source of information for the medical community. Signals can be grouped into three types: periodic, static, and random.

Periodic signals (sometimes called repetitive signals) are symmetric and repetitive signals. An ECG, for example, has a very predicable pattern (unlike an EEG, which has a variable signal). Other examples of periodic signals are blood pressure waveforms and respiratory waveforms. Periodic waveforms are commonly monitored, and their frequency is counted (and recorded or displayed). Periodic waveforms are shown in Figure 4.3.

Static signals are very regular and unchanging. Temperature is a good example; it might change, but it does so very slowly and

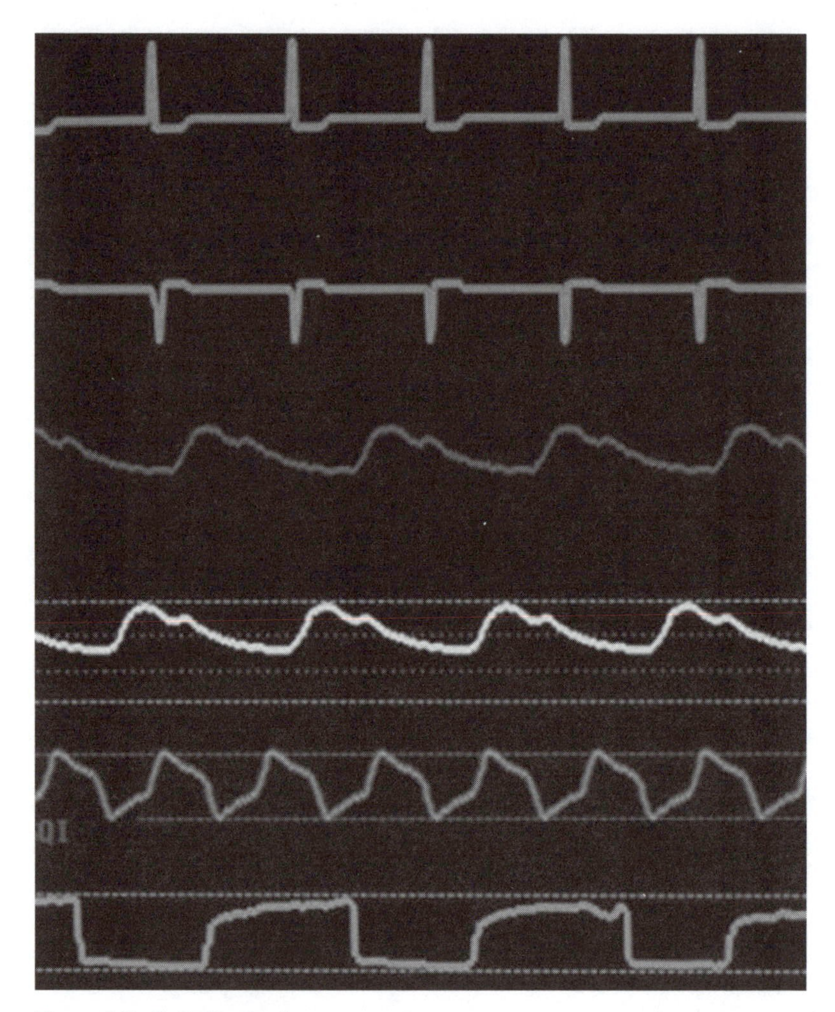

Figure 4.3. Periodic signals.

in very small increments. These signals often appear on a display as a straight line.

Random signals seemingly have no pattern, although some signals (like an EEG) may have some common characteristics. Random signals may have a common frequency but still have an

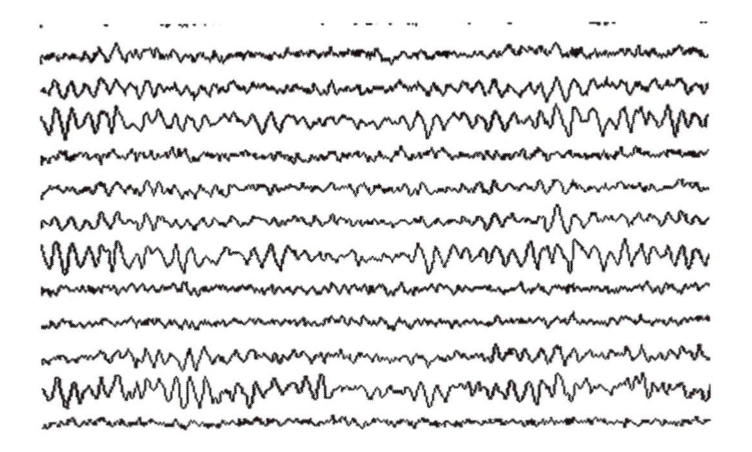

Figure 4.4. Random signals.

unpredictable waveform. Electromyograms, which record the electrical activity of muscles, also fit this type of signal. Random waveforms are shown in Figure 4.4.

Signal conversion

Most (if not all) physiological signals are analog; most (if not all) monitors and medical devices are digital. Because of the differing signals, conversions are necessary to coordinate communication between the patient, the sensors, and the signal processing systems.

Analog-to-digital conversion: A varying, analog signal from the patient is converted into ones and zeros in a process that is often called A/D conversion. There are many ways to perform this conversion. An illustration of this process is shown in Figure 4.5.

Digital-to-analog conversion: A digital signal of ones and zeros is converted into an analog signal in a process that is often called

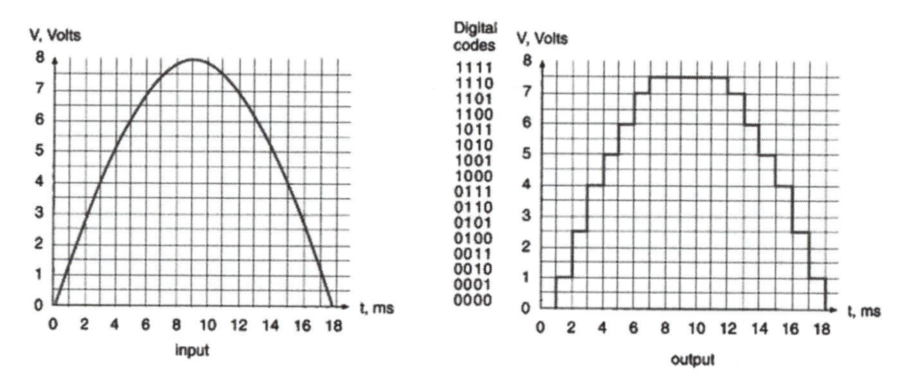

Figure 4.5. Analog-to-digital conversion.

D/A conversion. This may occur when a display of a waveform is created. Again, there are many methods to perform this.

Signal noise

Signal noise, or interference with the monitoring of human electrical signals, can hinder accurate patient observations. What are the sources of signal noise?

- **Lights** (especially the 60-hertz (Hz) "hum" from flourescent fixtures) – All electricity is received from the power company at a frequency of 60 Hz.
- **Human body** – The many signals within the body operate at different voltages and frequencies. These systems include the brain, the heart, the muscles, and the digestive system.
- **Computers** and their cables – These can add signal noise and electromagnetic interference.
- **Other machines being used** – Those machines that generate electrical energy or have motors (such as electrosurgical units in the operating room) are a common source of interference.

Signal conditioning

The **instrumentation amplifier** is the backbone for most medical instrumentation. It matches impedance and provides amplification and signal conditioning for physiological signals from the patient.

The **isolation amplifier** is basically a safety device. Because the patient needs to be protected from the monitoring circuitry and yet the physiological signal needs to be evaluated, the isolation amplifier was developed. Many techniques are available to electrically isolate the equipment from the patient; however, most systems use an optical link.

A note about circuitry: The fundamentals of circuit layout and troubleshooting were essential years ago. Good BMETs would evaluate component performance and replace individual items when necessary. As technology has evolved, there has been a shift in emphasis from components to circuit boards. In general, this shift has diminished the time spent analyzing common circuits or the ability to replace malfunctioning components such as capacitors or transistors.

STUDY QUESTIONS

1. Define *in vivo* monitoring in humans. Provide an example.
2. What is the fundamental purpose of sensors when used in patient care?
3. Briefly characterize known sources of error in sensor systems.
4. Describe the components of a surface electrode and the purpose of each part.
5. Describe some of the challenges of the connection between the surface electrode and human body.

6. Compare and contrast surface electrodes and needle electrodes.
7. Compare and contrast periodic signals and random signals and give examples of each.
8. Identify and describe some sources of noise in human monitoring.

FOR FURTHER EXPLORATION

1. The clinical setting can be a tough place for sensors. Patients are not always docile; devices must be thoroughly cleaned; and dropping, banging, and smashing devices are a secondary result of the fast-paced environment of patient care. How does this influence the design of sensors? How can BMETs balance the needs of staff to care for patients and the need to preserve technology? Examine the sensors described in this chapter and evaluate their durability and appropriateness for the clinical setting.
2. What are some of the significant advantages of using LEDs as indicators in medical devices? Discuss their durability, life span, and electrical requirements.
3. A number of devices employ human physiological signals but may not be commonly used in patient care. For example, many laptops have the capability to use biometrics to identify the user. Some gaming consoles use sensors to detect player motion. Explore some of these sensors used to interface humans to devices (not necessarily medical applications). Describe these sensors. Use the Internet to research and document your answers.
4. Researchers who investigate the human body using animal experimentation often employ unique sensors to measure a

wide variety of variables. Use the Internet to explore animal sensors. Select several and describe how they are made and what they measure. How do they differ from those found in the clinical setting?

5. Biofeedback is a common practice using human signals converted to sound. Explore this process to understand how the mind can be trained to control physiological signals. Identify the signals commonly used in this technique. Describe this process and its believed benefits to people who use it.
6. Research some human physiological signals. Describe the signals and how they may be measured. Identify whether the signal is electrical, mechanical, or some other type.
7. Many of the devices described in this book use sensors to extract physiological data from patients. Research devices described in this book on manufacturers' Web sites. Describe which type of sensor is used and the type of data retrieved. Be specific about the sensor, identifying the electronics or mechanical system involved.
8. Use the Internet to explore an instrumentation amplifier circuit. Obtain a simple circuit diagram. Identify and describe the basic functions of the parts of the circuit. Follow the physiological signal through the circuit.
9. The ability to regulate our core body temperature can be compromised by many situations, conditions, and diseases. List and describe three such conditions. Infant prematurity can limit the ability to control body temperature. Research the relationship between prematurity and temperature regulation. Describe techniques that are used to assist newborn patients with temperature regulation.

5

The heart

LEARNING OBJECTIVES

1. list and describe the purpose of patient monitoring
2. list and describe the characteristics of the ECG electrical waveform
3. describe cardiac events such as MI and PVC
4. list and describe the differences between the 3-lead and 12-lead ECG
5. list and describe the many names for the technique of blood oxygen saturation measurement
6. describe the reasons that pulse oximetry monitoring is commonly used
7. describe how pulse oximetry works

Introduction

The beat of the heart, including the electrical signal transmitted to create the beat, is one of the most familiar waveforms found in health care. It deserves this status. The electrical activity of the heart is fairly simple to measure and tells a great deal about the health of the patient. In the very early 1900s, Willem Einthoven won the Nobel Prize in medicine for his work identifying and recording the parts of the electrocardiogram.

Patient monitoring

Monitoring a patient's heart is a very common procedure. A patient's **vital signs** are often tracked at regular time intervals or continuously (these are usually blood pressure and pulse – both related to heart function). Technology can be used to track the patient's condition without someone standing next to the bedside. In addition to tracking the condition of a patient continuously, monitoring can provide diagnostic information upon which treatment decisions are based. Vital signs such as blood pressure and respiration are discussed in their respective chapters.

Note about assumptions for vital signs: Most patients in most hospitals are adults (they are simply sicker more often). Therefore, when a vital sign is listed and *unlabeled* (as to the age/type of patient) it is *assumed* to belong to an adult.

Electrical activity of the heart

The heart generates **electrical signals** to contract (pump blood). Cells within the heart have the ability to depolarize and repolarize, and this process generates electrical energy. The electrical signal in the heart begins at the sinoatrial (SA) node, which stimulates the **atria** (the two upper chambers of the heart) to contract. The signal then travels to the atrioventricular (AV) node, followed by the Bundle of His, and then through the Purkinje network, which stimulates the **ventricles** (the two lower chambers of the heart) to contract. When the ventricles are being stimulated to contract, the atria are repolarizing.

The electrical signal is called an **electrocardiogram**. The abbreviation used for this can be **EKG** (which comes from German *Elektrokardiogramm*) or **ECG** (you will find these acronyms used interchangeably). Some people prefer EKG since it is unlikely to be confused with the EEG when spoken. There is no consensus on the label EKG or ECG. There may be times when the terms are used together. As an example, an ECG monitor can display a 12-lead EKG. Called either EKG or ECG, the electrical waveform has a patterned shape, and its peaks are labeled with letters, as shown in Figure 5.1. Try to visualize the electrocardiogram pattern as a reflection of the motion (contracting and relaxing) of the heart.

The first little upward notch of the ECG waveform is called the **P wave**. The P wave indicates that the atria are contracting to pump blood. The next part of the ECG is a short downward section connected to a tall upward section. This part is called the **QRS complex**. When the Bundle of His fires, the ventricles contract to pump blood. The large amplitude of voltage is required because the ventricles are the most muscular and dense part of the heart. The next upward curve is called the **T wave**. The T wave indicates the resting period (repolarization) of the ventricles.

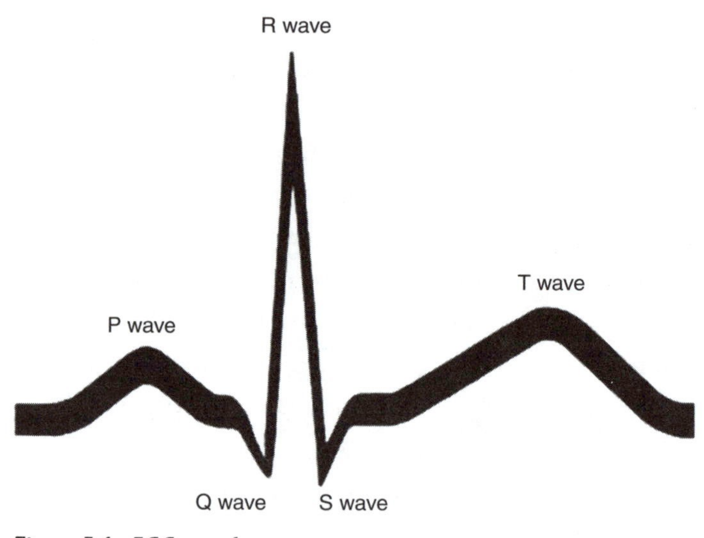

Figure 5.1. ECG waveform.

The number of waves per minute is the **heart rate**. Notice that heart rate is measured per minute rather than per second as most other electrical signals.

The amplitude of the QRS segment can be measured at the skin and is about 1 millivolt (mV).

ECG monitoring

ECGs can provide information about the condition of the patient. Some physiological events that a medical team might look for include:

- Fibrillation – the fluttering of the heart, which is essentially randomized electrical signals that result in chaotic and ineffective contractions.

- Asystole – no electrical activity of the heart, "flat line." Patients in this condition are clinically dead.
- Rhythm disturbances – includes premature ventricular contractions (PVCs), where the ventricles contract at the wrong time.
- Conduction abnormalities (disruption of the electrical pathways of the heart).
- Size of the heart chambers (enlargement or atrophy).
- Position of the heart in the chest (axis).
- Rate at which the heart is contracting (*bradycardia* – too slow, *tachycardia* – too fast).
- Diagnosis of myocardial infarction (MI) (heart attack).
- Ischemia (lack of blood flow to the heart) and other disease states.
- Effect of cardiovascular drugs.
- In exercise testing, effects of physical workload upon the heart.

It is important to know what the ECG *should* look like because it is often the job of the BMET to evaluate the performance of heart monitors. However, the study of the subtle waveform changes that occur in disease conditions, although vital to the medical team, is not required of BMETs. Therefore, a basic understanding of the general ECG waveform characteristics is very important; however, in-depth study of abnormalities is not required of BMETs.

Cardiac monitors typically display the ECG waveform as well as a numerical display of the heart rate (beats per minute). In addition, monitors have alarms that can be set to notify staff if minimum and/or maximum heart rates are exceeded. This data may also be networked to a central station. Typically located at the front desk of a nursing unit, a monitor can display the ECG waveforms of many patients.

ECG waveforms are most often obtained from *surface electrodes* stuck onto a patient's skin. It is important to note that waveforms can also be obtained in other ways. One common example is to monitor a fetus during delivery. An electrode wire is connected to the skin of the head. The needle electrodes are explained in Chapter 4.

Throughout electrical training, the terms "lead" and "wire" are used interchangeably. This is not true for cardiologists and other clinicians. They define a **lead** as a *view* of the electrical activity of the heart from a particular angle across the body. The wires connected to the patients are paired, in different combinations, to obtain the many views needed for clinical interpretation. Since voltage needs two connecting points, combinations of the various connecting points (wires) provide clinicians with leads or *views*.

Three-lead ECG monitoring

Only two connections are required to measure the electrical activity of the heart. For example, a very basic ECG can readily be seen by holding a lead in each hand and using some simple signal processing amplifiers. However, the importance of a diagnostic-quality signal requires more attention to the patient connections.

There are three connections to the patient in a **three-lead ECG**. The connections are most often labeled RA (right arm), LA (left arm), and LL (left leg). The labels may be a bit confusing because the electrodes are often placed on the torso not the extremities. The arm connections are usually placed near the shoulders and the LL is placed near the bottom of the rib cage (patient's left side). Each lead graphically shows the voltage (potential difference) between two of the electrodes. A three-lead

ECG illustrates the electrical activity of the heart in three different ways using the three electrical wires connected to the patient:

Lead I – LA (positive) and RA (negative) (memory aid: one L in the configuration)

Lead II – LL (positive) and RA (negative) (memory aid: two Ls in the configuration)

Lead III – LL (positive) and LA (negative) (memory aid: three Ls in the configuration)

One electrode is not used as a part of each of the lead selections. It is used as the "ground" reference lead. That is, if you are monitoring lead II (RA and LL), the LA electrode becomes the ground. This combination comes from the work of Willem Eithoven and is called Eithoven's triangle. The leads, called bipolar limb leads, form this triangle, which is illustrated in Figure 5.2.

Three-lead monitoring is the most common type of patient connection to look at the electrical activity of the heart. When constant monitoring is required, this method will be used.

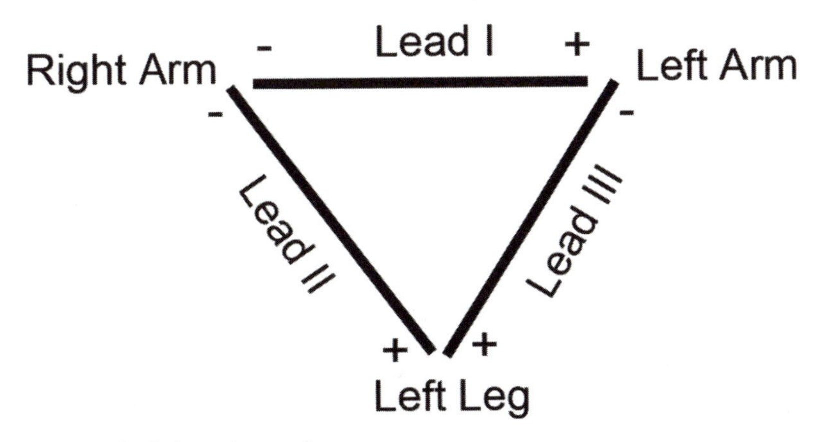

Figure 5.2. Eithoven's triangle.

However, a three-lead ECG does not give as much diagnostic information as possible.

Five-lead ECG monitoring

The **five-lead** ECG provides *six views* of the heart. Three of the views are the same as the three-lead ECG - leads I, II, and III - plus three additional views called aV_R, aV_L, and aV_F. These additional leads are called **augmented leads**. More views provide the medical team with more information with respect to heart tissue. There are usually five patient connections, three connections in the same locations as for the three-lead ECG. There are the same two shoulder-area electrodes, LA and RA. The LL electrode is in the same location, near the lower edge of the rib cage. Two additional electrodes are added to the patient: one below the rib cage on the right side (RL) and one close to the center of the chest (Chest). RL is always the "ground" reference lead.

12-lead ECG monitoring

A 12-lead ECG monitor uses *ten electrodes*, four attached to the limbs (although often on the torso) and six connected in a line on the chest wall. These connections to the patient are called unipolar leads, can provide extensive information, and are commonly used for diagnostic purposes. The electrode attached to the right leg serves as a "ground." The electrodes can obtain 12 different voltages of the heart by making 12 comparisons. Accurate lead *placement* is extremely important. The views that clinicians look at are I, II, III, aV_R, aV_L, aV_F, V_1, V_2, V_3, V_4, V_5, and V_6. Note that the 12-lead ECG provides the six views from the five-lead ECG plus six additional views. The addition of more electrodes attached to the patient's body expands the information from two dimensions into three dimensions. The diagnostic use of the additional leads is not within the scope of this book; however, for cardiac patients, the additional information is very important.

Twelve-lead ECGs are usually done for diagnostic purposes, and a patient typically is not connected to all ten electrodes all the time. Twelve-lead tracings may be useful during stress testing (when the patient is constantly monitored during exercise or activity). This provides the medical team with a great deal of cardiac information.

Lead identification

To help clinicians properly place the electrical connections, color-coding is used. In the United States,

RA connector has white snaps or wires

LA connector has black snaps or wires

LL connector has red snaps or wires

RL connector has green snaps or wires

Chest connector has brown snaps or wires

Clinicians use the following mnemonic device to remember where to place the colored connectors: "Right is white" to remember to place the white lead on the (patient's) right-side shoulder. "Black is opposite of white" so black goes on the left shoulder. Then "smoke over fire" reminds the clinician to place the red lead under the black lead. "Green under white" indicates that the green lead is placed on the right leg, and the brown lead is often placed in the V_1 position (near the center of the chest).

As a BMET, understanding where connections are made can be very important when assessing a reported problem with an

ECG monitor. Often a problem lies with the connectors, electrodes, and patient connections rather than the monitor. Evaluating electrode placement and condition can serve as a starting place for the assessment of difficulties, problem resolution, and communication with the clinical staff.

Long-term ECG recording

Long-term ECG recording, also called **Holter monitoring,** can detect ECG abnormalities that may occur over time. The patient wears ECG electrodes that are connected to a recording device (originally, Holter monitors were cassette tape recorders, but now they are digital recorders). The recording device is quite small and is shown in Figure 5.3. Patients wear these monitors

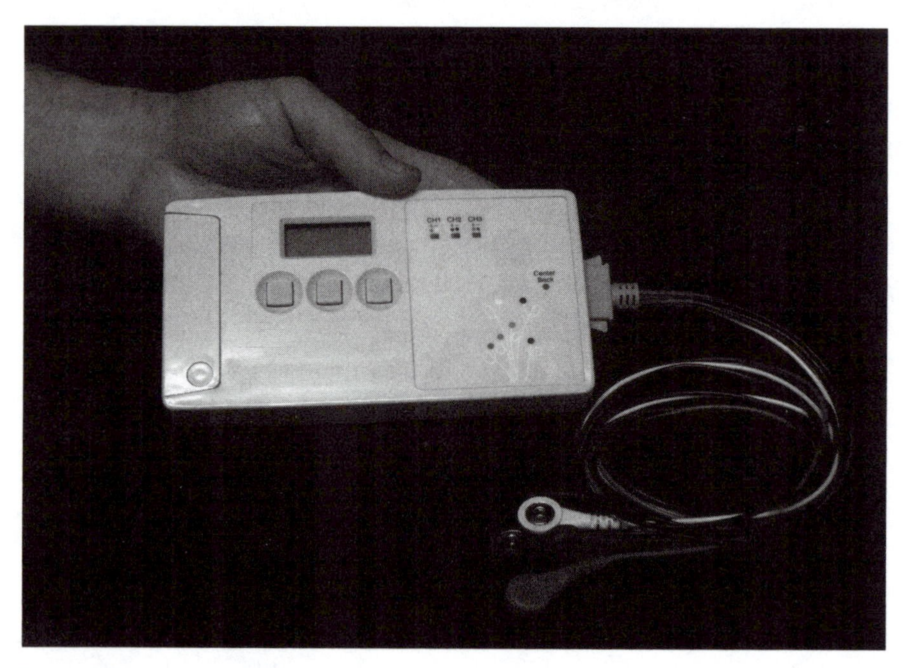

Figure 5.3. Long-term ECG recorder.

as they go about their normal activities. Typically these recordings are done for 24–48 hours.

Fetal heart monitoring

Heart monitoring of infants in the uterus is a common procedure. In a routine examination, a hand-held Doppler device is used. This uses ultrasound to detect fetal heart activity. Using ultrasound allows the fetal heart rate to be detected even with the mother's heart rate activity present. It also can be converted to sound to "hear" the heart beat. Observed heart rates are usually 120–160 beats per minute. During labor, continuous monitoring is preferred. The electrical activity of the fetus is monitored for signs of stress and labor complications. Additional information and a photo of a device used to monitor fetal heart rates during labor are in Chapter 15.

During vaginal delivery, the heart rate of infants can be electronically measured directly by placing a wire into the scalp. This is *internal fetal monitoring*. Needle electrodes were discussed in Chapter 4.

Pulse oximetry

Also called SaO_2, SpO_2, pulse ox, and sat or sats (short for saturation), pulse oximetry is a simple, noninvasive method of monitoring the percentage of hemoglobin (Hb) in the blood that is saturated with oxygen. Sensors connected to the patients contain two parts – LEDs and photodetectors. Red and infrared LEDs are used as light sources. The **red light** has a wavelength of

660 nanometers (nm), and the **infrared light** has a wavelength of 940 nm. These light wavelengths are passed through the skin, often a finger tip, ear lobe, or foot (premature infants). Hemoglobin in the blood both absorbs and reflects these wavelengths, depending on the amount of oxygen that is being carried. Photodetectors measure the resulting intensities of light in the two wavelengths. The color of the blood changes with the amount of oxygen in it. As the blood is pulsed past the light sources, the resulting signal varies – the blood flow creates a pulsating signal that is directly related to the heart rate/pulse rate.

Note: Pulse oximetry readings do not indicate respiration rate. While the respiration rate may influence oxygen saturation, the number of breaths per minute is not displayed during pulse oximetry monitoring.

This type of monitoring also shows the changes in saturation as the heart beats. Because of this, pulse oximetry will indicate *heart rate.* In many cases, basic stable patients are monitored using pulse oximetry. The sensor is very easy to apply; there are no sticky pads as required for ECG monitoring. A reusable pulse oximetry probe is shown in Figure 5.4. Patients do need to remove clothing or be awakened when the sensor is applied. In addition, only one connection is needed. Lastly, the device itself is very simple and relatively inexpensive. Heart rate determined using pulse oximetry requires no counting (as required for the measurement of pulse rate by counting pulsations at the wrist), which can introduce human error.

In addition to the simplicity of heart rate measurements, pulse oximetry is a very simple method to evaluate the quality of **respiration activity** (versus counting breaths, which may or may not be successful in oxygenating blood). The quality of respiration can be determined from the percentage oxygenation value reported by

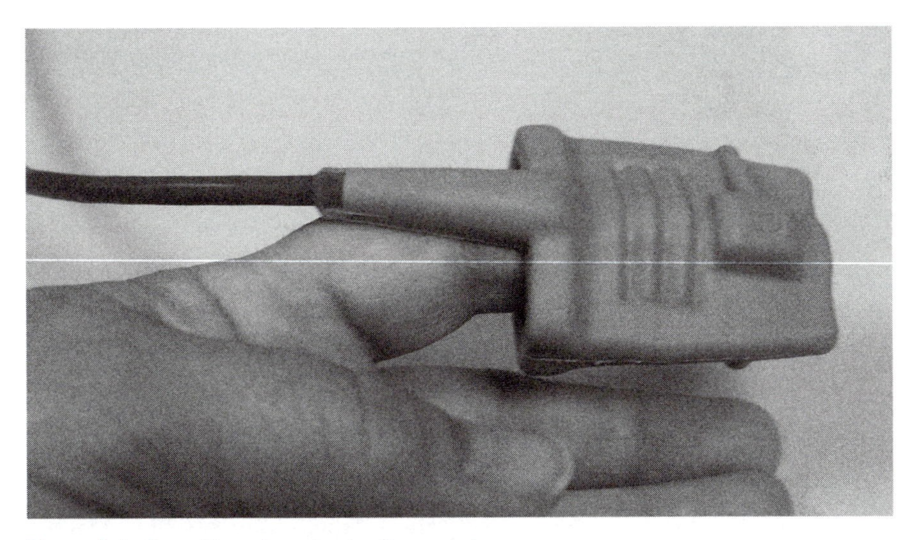

Figure 5.4. Reusable pulse oximetry finger probe.

these devices. Again, note that pulse oximetry does not measure respiration rate. The device counts and displays heart rate and a percentage value that indicates the percentage of hemoglobin (Hb) saturated with oxygen.

Most healthy people show a pulse oximetry of 95% or above (indicating 95% of the hemoglobin is carrying oxygen). A *reading lower than 90%* may be due to any factor that affects blood, hemoglobin, and oxygen circulation in the body. Lower readings may accurately reflect a patient problem or some situation that prevents accurate saturation measurements. These may include:

- Excessive bleeding
- Lung problems, such as pneumonia
- Cigarette smoking
- Blood vessel problems
- Respiratory disease or chronic obstructive pulmonary disease (COPD)

- Stress or pain
- Hypothermia
- Nail polish

Perhaps one of the best features of this type of monitoring is that there is constant awareness of patient condition. Patients who experience difficulties in effective respiration show changes in the saturation readings immediately. In addition, changes in heart rate are immediately apparent.

Movement of the patient can be a problem in producing accurate pulse oximetry readings. Movement might include something as simple as shivering. A technique called Masimo SET (Signal Extraction Technology) is able to better reduce the inaccuracy caused by movement.

Many types of devices monitor pulse oximetry. For example, this technique is often built into physiological monitors that include blood pressure and temperature measuring devices. There are a wide variety of types of **probe**, the sensor that is attached to the patient. Some are reusable and some are disposable. Usually, these are made to be quite durable.

STUDY QUESTIONS

1. What variables that represent a patient's vital signs are commonly measured?
2. Make a table. In the first column, list the letters that represent the parts of the ECG (P, Q, R, S, T), and in the second column, describe the motion/action of the heart associated with that point of the electrical signal.

3. Identify and describe the clinical difference between lead and view.
4. What is the clinical benefit of using more than two patient connections to record the electrical activity of the heart?
5. Predict the resulting lead I waveform in a three-lead ECG if the RA is the positive connection and the LA is the negative connection. Use your knowledge of electrical connections to predict the result.
6. What are the augmented ECG leads?
7. Sketch a human torso. Label the patient's right and left sides. Illustrate the locations of the electrical connections for a five-lead ECG. Label each connection with the associated color.
8. Describe the patient information provided by pulse oximetry monitoring.
9. Compare respiration rate data and pulse oximetry data. How are they related to the activity of the lungs?

FOR FURTHER EXPLORATION

1. The ECG waveform is commonly used as a symbol of health care. Locate examples of stylized representations used in marketing. Why is this waveform such a popular symbol?
2. Convert the heart rate vital signs information in Table 5.1 to hertz. Describe why it is uncommon to measure and record human ECG data in hertz.
3. Research a common heart attack. Describe the underlying physiological changes that produce cardiac difficulties. Identify the risk factors for a heart attack. Use the Internet to document your answers.

TABLE 5.1. *Heart rate ranges*

Vital sign	Infant	Toddler	School-aged child	Adult
Pulse rate per minute (heart rate) – values are listed in a range for normal	120–160	90–140	75–100	50–90

4. The beeping heard in television shows is often a media attempt to represent each pulse of a QRS complex. Why has the rhythmic beeping become a symbol of life in the media? Identify hospital scenes where beeping is heard but no cardiac leads are attached to the patient.
5. Patients who experience fibrillation do not have the straight-line (flat-line) waveform typically shown in the media. What does a fibrillating heart waveform look like? Why does the media represent it a different way?
6. Explore and describe a stress test. What does this entail? What type of monitoring is done during this test? What type of patient has a stress test?
7. Suppose a clinician explains that a three-lead ECG monitor isn't working. Lead III displays nicely but the other leads I and II are very noisy – what could be the problem? Is it likely a problem with the device?
8. The wires for the various locations on the patient have standard colors. Draw a 12-lead ECG configuration, identifying the colors used in the various locations.
9. A common source of ECG monitoring problems occurs in part of the patient cable, which connects the patient and the monitor. A breakage of wire (under the insulation) occurs and is not visible upon inspection. How could a BMET determine if the patient cable has a fault? Why are breakages so common?

10. Explore this Welch Allyn booklet (http://www.iupui.edu/~bmet/book/ECG_basics.pdf) that explores ECG recordings in depth. Summarize the major points.

11. Research electrical connections to measure ECG and the term unipolar lead. Why are these connections termed unipolar? What is the reference point for each augmented lead?

6

Cardiac assist devices

LEARNING OBJECTIVES

1. describe defibrillator waveforms and characteristics
2. define the term AED and describe why it is useful
3. describe cardioversion
4. describe assistive technologies such as LVAD, pacemaker, implantable defibrillator, heart-lung bypass, ECMO, cardiac catheter, and IABP

Introduction

Technology is used to support heart function in numerous ways. Perhaps the most common and well-known device is the defibrillator. In addition, the amazing life-saving powers of artificial hearts and implantable devices capture the attention of the public and news media on occasion.

ACLS (advanced cardiac life support) is a detailed protocol to provide life-saving cardiac care. BMETs may be involved in the selection of technology and technical protocols used in a crisis. In addition, BMETs may help establish training programs and awareness for the technology involved in emergency response.

Defibrillators

A **defibrillator,** which is often used in life-saving moments on television medical shows, is a fairly simple device that delivers a large amount of energy across a patient's chest when the heart is not beating properly. When the heart is not beating with a pattern, all cells are contracting at different times; this is known as **fibrillation.** Defibs (as they are often called) in use today send about 3,000 volts (V) - up to 20 amperes (A) - across a patient's chest. (*Note:* The amount of voltage and current vary tremendously based on the patient's size and therapeutic settings.) This large amount of energy causes all cells in the heart to depolarize at once. This action may allow the SA node to resume function and generate a normal electrical pattern. Sometimes the SA node can recover, but sometimes it cannot. The shape of the energy waveform has evolved with experience. The **Lown waveform** was

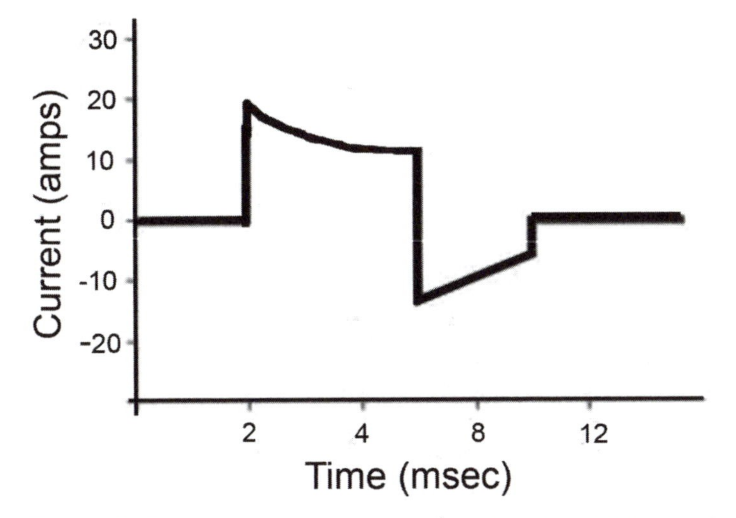

Figure 6.1. Biphasic defibrillator waveform (produces about 150 Joules of energy).

the therapeutic waveform standard until recently. It looks like a damped sine wave.

The **biphasic waveform** is the standard of care in defibrillators today. It has both a positive and a negative component and produces much better results more quickly. In addition, biphasic waveforms are able to deliver results with less energy. The waveform is approximated in Figure 6.1.

A defibrillator is connected to a patient in a specific way. The placement of the paddles or sticky pads on the chest delivers the biphasic waveform in a way that mimics normal cardiac electrical direction. One paddle is placed on the right side of the patient, near the clavicle. This paddle is often labeled "sternum." The other paddle is placed on the left side of the patient near the ribs, below the breast. This paddle is often labeled "apex." This configuration is often marked on the paddles (or sticky pads).

The proper functioning of a defibrillator is vital to successful patient care when a patient is experiencing cardiac problems.

BMETs ensure adequate testing procedures (including when the device is not connected to an electrical outlet and is running on battery power) and work to establish performance assurance protocols. BMETs also measure the power output of defibs to ensure that the device settings match the actual delivered energy. (Too little energy can fail to restart the heart, and too much electrical energy can injure the patient.) Test equipment usually includes a simulated human body load.

Even though defibrillators are excellent devices for restoring cardiac electrical activity, they depend on the administration of energy in a very timely way. A general time frame suggests that for every minute in fibrillation, a patient drops approximately 10% in survival rate. Therefore, improving the time between the onset of fibrillation and the application of the biphasic shockwave is a critical goal.

Cardioversion is the process of shocking a heart at exactly the right point in an ECG waveform. Cardioversion is required when a heart is functioning somewhat and is not in complete fibrillation. The defibrillator is capable of analyzing the ECG waveform from the patient to deliver the shock at the proper time. The shock is delivered immediately after the R wave peak. Essentially the physician applies the paddles to the patient, pushes the shock button and nothing happens until the defibrillator can deliver the energy at the right time. Cardioversion is a commonly tested function to verify proper device performance. Specialized defibrillator test equipment can simulate the ECG waveform needing cardioversion.

AEDs – automated external defibs – are currently in the news because these devices are now available in a wide variety of places including in planes, in police cars, and with security personnel. In O'Hare Airport in Chicago, they hang on the wall in glass cases (like fire extinguishers). The airport decided that there should be a defib within a one-minute fast walk from anywhere in the airport. Basically, AEDs are the same as defibs, but they generally have no waveform display, use stick-on patient pads rather than paddles, and calculate the power to be delivered automatically. An important feature of AEDs is the patient pads that check for ECG signals so that a patient who is *not* in fibrillation will *not* be shocked. Many of these units provide audible instructions so that they can be operated by someone without medical training. In addition, the devices have memory that stores the ECG waveforms of the patient and makes them available to medical staff when the patient is transported to a hospital. The American Red Cross has a wealth of information about AEDs. Perhaps you have seen advertisements for AEDs for people to keep in their homes, reserved for emergencies, like fire extinguishers. One type of AED is shown in Figure 6.2.

Ultimately, the goal of making AEDs widely available is to improve the survivability of cardiac fibrillation victims by decreasing the time until defibrillation can be applied. The AEDs themselves must be self-diagnostic since testing is expected to be performed by untrained people without test equipment.

Artificial hearts

Implantable hearts like the Jarvik 7 are unusual and are not the typical choice for patients with heart problems, even though

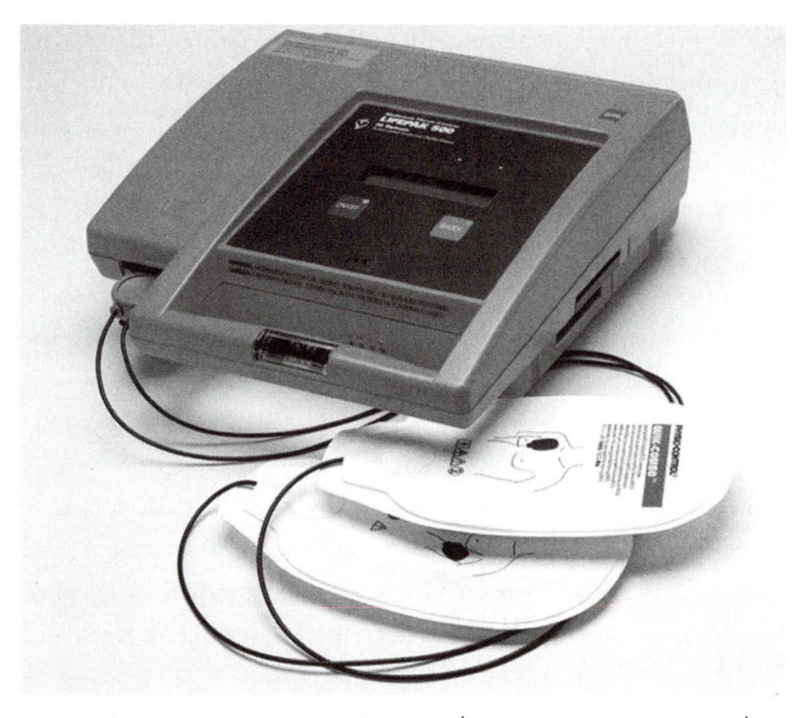

Figure 6.2. Automated external defibrillator. (Photo courtesy of Medtronic.)

their use is often highly visible in the press. Complete heart replacement devices are often not the responsibility of most BMETs except those with highly specialized training.

External temporary cardiac assist devices are used when the patient has had significant damage to the heart muscle. Blood is diverted to an external ventricle giving the heart time to heal. These are often used on a short-term basis when the patient is not responding to other treatments and a transplant is necessary. *Ventricle assist devices (VADs)* and *left ventricle assist devices (LVADs)* can be extremely costly and carry many associated risks.

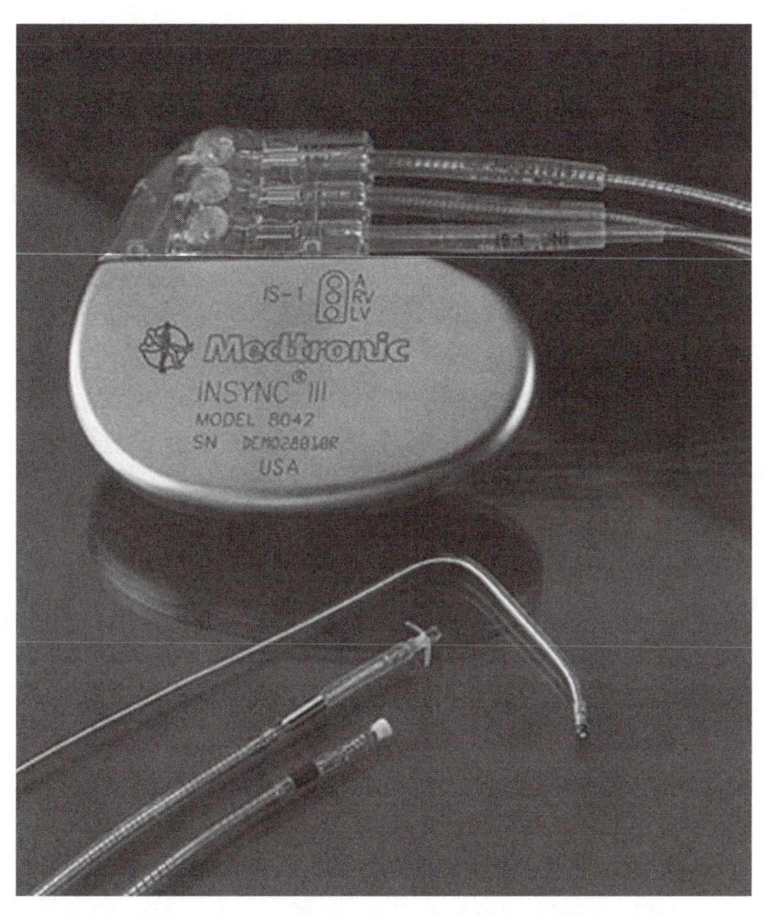

Figure 6.3. Implantable pacemaker (shown with leads). (Photo courtesy of Medtronic.)

Pacemakers

Pacemakers can support the proper function of the heart to keep regular electrical rhythm, or they can stimulate the heart only when it does not beat properly. An example of a pacemaker is shown in Figure 6.3. In general, only BMETs with specialized training are involved in supporting pacemakers. There are several types of pacemakers.

- **Transcutaneous external cardiac pacing or transthoracic external pacing (TEP)** uses an external type of pacemaker that uses electronic pads placed on the skin of the patient. TEP is often done in emergencies and for temporary measures until an implantable pacemaker can be provided. A blend of time of the pulse and amount of energy delivered is balanced to find good cardiac effect and yet make the process tolerable for patients. Synchronous pacing is often used to supplement the heart's own electrical activity.
- **Internal pacemakers** are often as sophisticated as small computers and are able to detect and correct electrical abnormalities. When the pacemaker is in place, the control circuitry and battery pack are located just under the skin below the clavicle. Wires are threaded through blood vessels into the correct cardiac position. These devices are able to respond to patient activity using ECG sensors on the case of the device. They can store patient data for download wirelessly through the skin. A variety of models assists patients with their different electrical conduction problems. Figure 6.3 shows a three-lead pacemaker, but two-lead pacemakers are also common.
- **External pacemakers** are connected to the patient's heart through internal wires but the electronics are located outside of the patient. This method of pacing is often used as a temporary measure (after surgery, for example) to assist the heart's own function.

Implantable defibrillators

Implantable defibrillators, also called implantable cardioverter-defibrillators (ICD): These devices deliver electrical

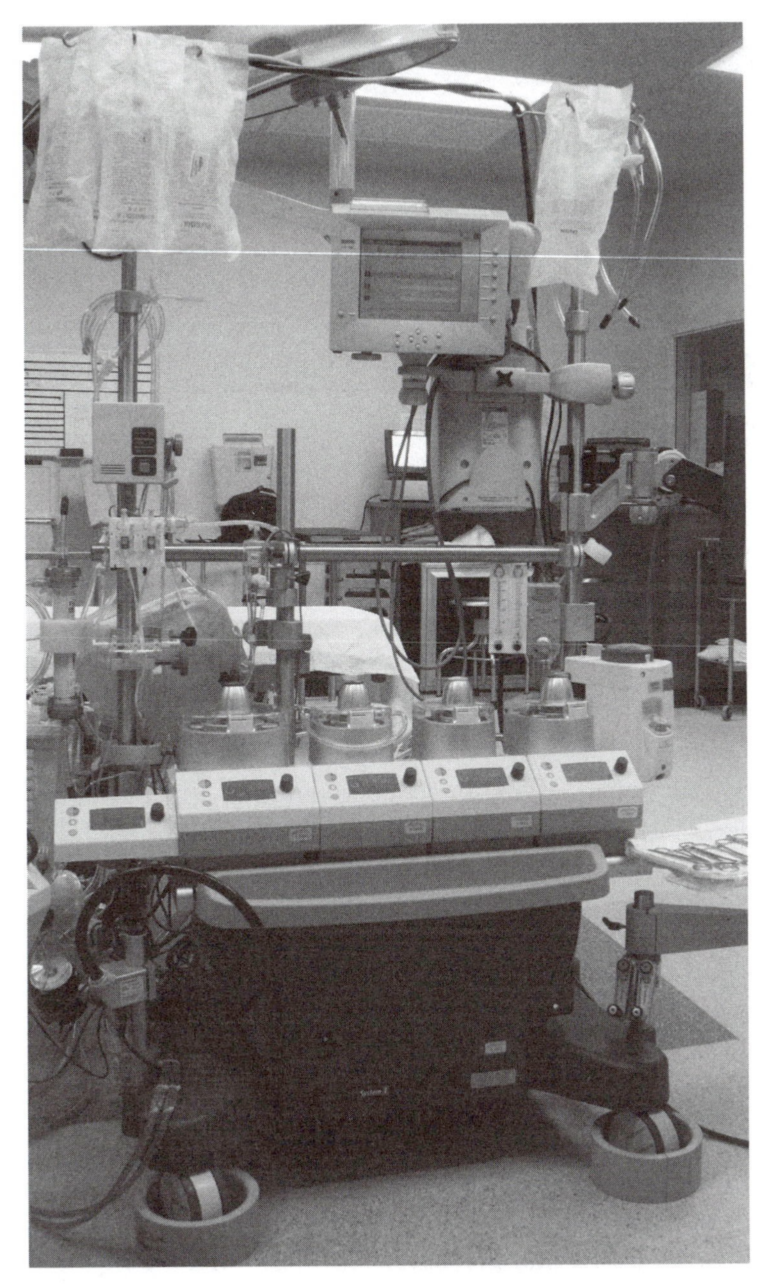

Figure 6.4. Heart–lung machine in an operating room.

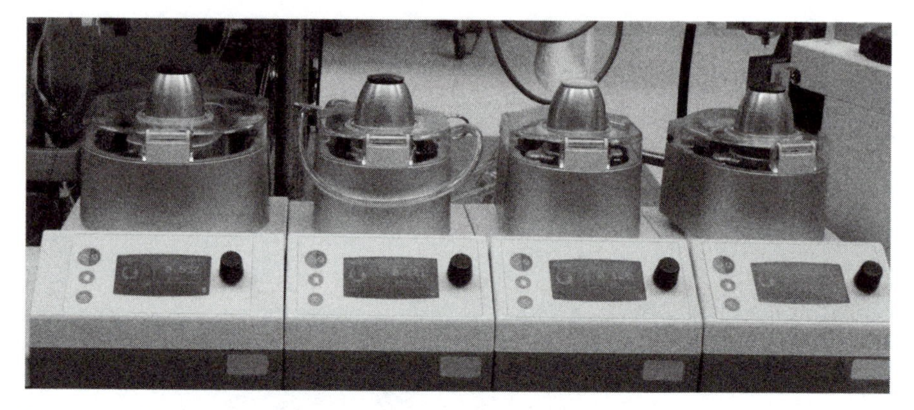

Figure 6.5. Heart-lung machine pumps.

shock to the heart when fibrillation is detected. They are similar to internal pacemakers in computer-like sophistication. However, they may not be needed to stimulate the heart very often or at all. While they look very much like pacemakers, their purpose is quite different. They act like an insurance policy to restart a heart if necessary.

Heart-lung machines: These devices are also called cardiopulmonary bypass and are commonly used during open heart surgery. The machine oxygenates and pumps the blood for a patient. To minimize the damage that pumps can have on blood cells and other blood components, *roller pumps* (which squeeze tubing to move contents) and *centrifugal pumps* (which use centrifugal force to move blood) are used. Blood is oxygenated, often using a membrane that is carefully designed for gas exchange. The device includes a blood warmer as well. Figure 6.4 shows a heart-lung machine, and Figure 6.5 is a close-up of the pumps needed to move the blood.

Extracorporeal membrane oxygenation (ECMO): Originally exclusively used in the operating room (for surgeons to work

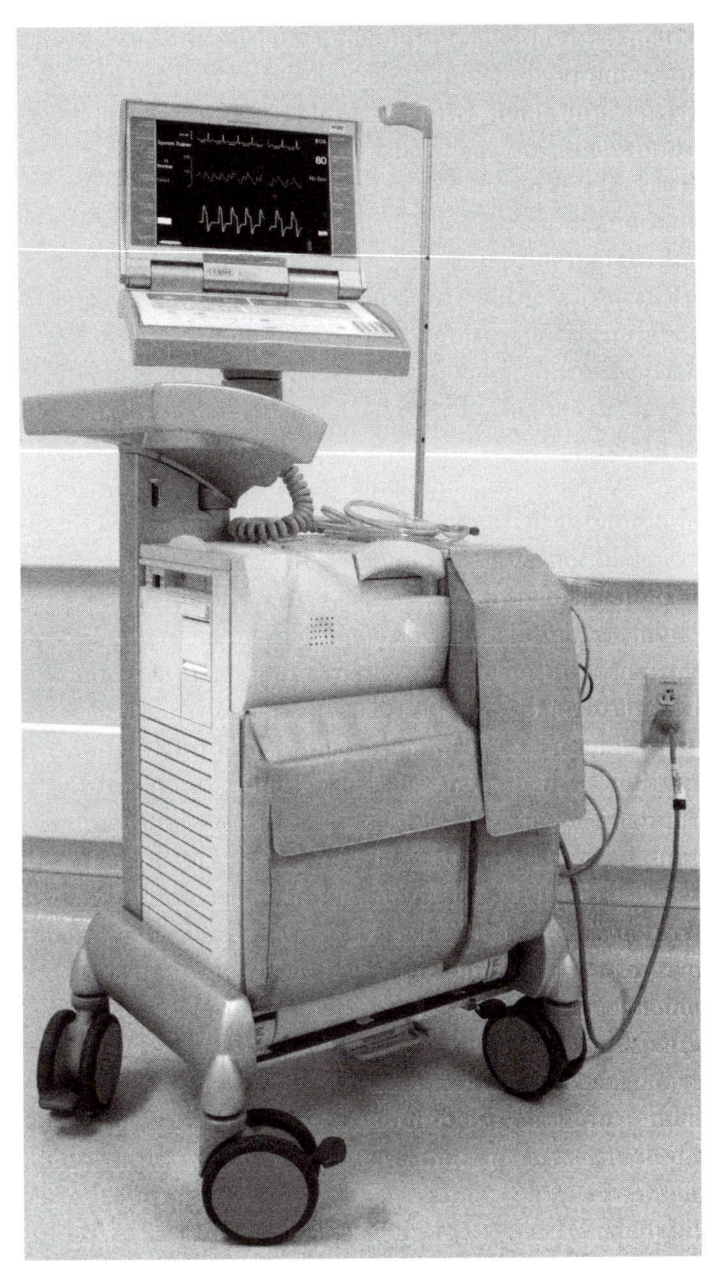

Figure 6.6. Balloon pump.

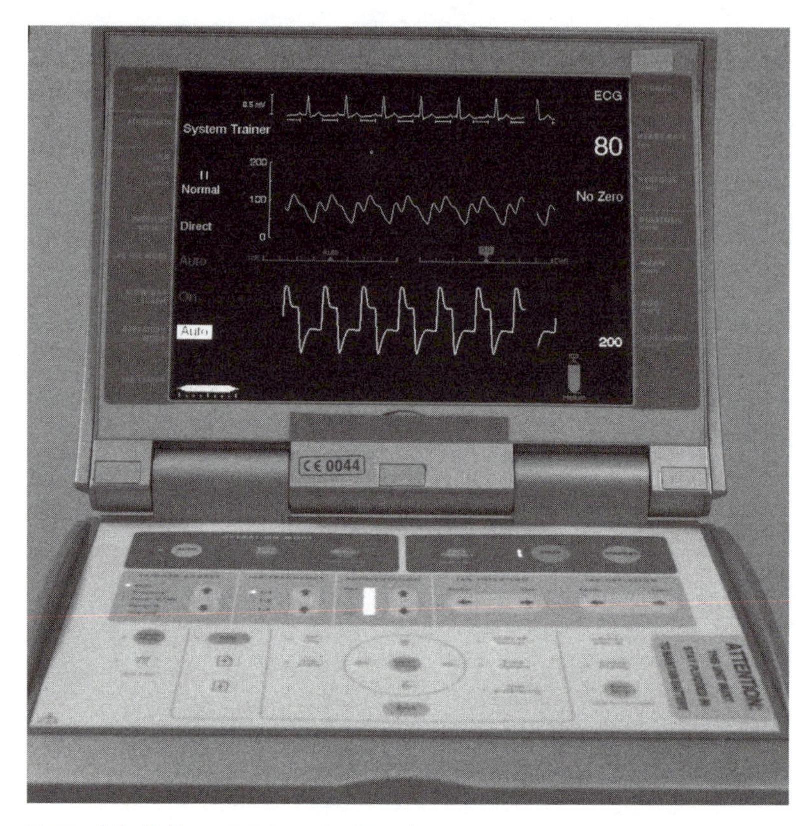

Figure 6.7. Balloon pump control panel.

on the heart), heart-lung machines have a related device called ECMO. This support is used on patients for longer periods (days) rather than hours during surgery. Another term for ECMO is extracorporeal life support (ECLS). Highly trained BMETs are involved in the support of these devices.

Valve replacements: Human heart valves are replaceable. Commonly replaced valves include the aortic valve and the mitral valve. Options for replacement include pig valves, which last between 7 and 15 years; mechanical valves, which are made of metal and plastic; and human donated valves.

Intraaortic Balloon Pumps (IABP): These devices are often used at the patient's bedside to assist the heart, allowing it to rest and recover, especially after a heart attack (MI) or cardiac surgery. A catheter with a balloon on the end is inserted into the femoral artery and threaded inside the aorta. The balloon is inflated and deflated, timed with the heart's rhythm, to improve coronary perfusion and reduce the cardiac workload. Balloon pumps use helium as the shuttle gas to inflate and deflate the balloon. The balloon inflates during diastole to increase pressure, deflates just before systole, which lowers pressure, making it easier for the heart to pump. The device is shown in Figure 6.6 and the control panel is shown in Figure 6.7.

STUDY QUESTIONS

1. Give an example of a patient condition that would require ACLS.
2. Compare the polarity of the electrical signal generated by the heart by sketching the shape of the heart, labeling the SA and AV node locations and then sketching the path of the electrical signal. Discuss and describe how the biphasic waveform (and the placement of the defib paddles) mimics this electrical conduction pathway.
3. Identify and define the conditions under which a defibrillator would be used on a patient.
4. Why is it vital for BMETs to regularly test the output of defibs? Why would careful calibration of the output energy be important?

5. Identify community locations where AEDs should be placed to ensure public safety.
6. Identify and describe the characteristics of patients who might receive implantable defibs.
7. Describe and define how IABP allows the heart to rest and heal.

FOR FURTHER EXPLORATION

1. Use the Internet to investigate the history of defibrillation. Describe historical milestones and include the point at which it became prevalent. How did defibrillation change the care of patients in cardiac arrest? Identify the latest major development.
2. Explore Web sites that describe AEDs. Characterize these devices. Describe the potential impact of prevalent and private ownership of these devices on the aging population. Police often carry AEDs in their cars since they often are the first responders. How has this changed the survivability of a fibrillating heart condition?
3. Use the Internet to explore the history of the artificial heart. Summarize this history including significant milestones. Who is Barney Clark? How did his implant impact artificial heart development? Is the use of a completely independent and implantable artificial heart in the near future? Use the Internet to research and document your opinion.
4. Heart transplant patients are often placed on LVAD in order to support their health while they wait for a suitable

donor. Describe the support functions of this device in this scenario. What restrictions do the patients have? How is the device powered? How long can a patient use a LVAD? What complications can occur with the use of an LVAD? Use the Internet to document and support your research.

5. Medtronic is a major manufacturer of implantable cardiac assist devices. Visit its Web site http://www.medtronic.com/tachy/patient/whatis_icd.html, which provides detailed information about ICDs. Summarize the function of an ICD.
6. Implanted devices of any kind can be influenced by the magnetic fields (emi), which are produced from motors, security equipment, and other devices. Summarize the information in this brochure, which discusses the impact of emi on implanted devices: http://www.medtronic.com/rhythms/downloads/icd_en 199300962cEN.pdf.
7. How common is the use of implanted pacemakers? What functions and sensors do pacemakers have in order to control a patient's heart with the utmost of flexibility? For example, can pacemaker patients run up a flight of stairs? Will their heart rate rise? What routine maintenance is required for patients with pacemakers? How long do the batteries in implanted pacemakers last?
8. The medical staff person who controls the heart-lung machine is called a perfusionist. What is his or her training? How might BMETs work with perfusionists to support the safe and effective use of heart-lung machines? Use the Internet to support your answers.
9. Describe the history of heart-lung machines. When were they first used and how did they work? How have they evolved?

10. Use the Internet to explore artificial heart valves. What are they made of? How long do they last? Why would a person need one? How has the design of valves evolved over time?
11. Describe the functions and components of an intraaortic balloon pump. Explore the cardiac benefits of IABPs. Why are they used? What therapeutic benefits do they bring to patients? Use the Internet to support your answers.

7

Blood pressure

LEARNING OBJECTIVES

1. describe the importance of blood pressure as an indicator of patient health
2. describe mean arterial blood pressure
3. describe the common methods of blood pressure measurement: manual reading using a sphygmomanometer; automated methods using NIBP devices; and direct arterial
4. describe the purpose and use of a Swan-Gantz catheter

Introduction

Blood pressure is an important indicator of performance of the human heart, lungs, and circulatory system. It is relatively easy to obtain, and readings can be taken at specific time intervals to track patient health. The peak and resting pressure of blood within the arteries is the most commonly measured blood pressure vital sign. Blood pressure does rise and fall as the heart beats and has a periodic waveform. The amount of pressure in the heart can reflect the overall health of a patient and can be an indicator of many diseases and conditions.

Indirect measurement methods

The most common **noninvasive blood pressure (NIBP)** measurement method involves a **blood pressure cuff** and a stethoscope (see Figure 7.1). The cuff is connected to a gauge that displays pressure in the cuff. The cuff and gauge together are termed a **sphygmomanometer.** The blood pressure cuff is usually placed around the arm and inflated to pressures displayed on the gauge. The manual technique involves a person who listens using the stethoscope to **Korotkoff sounds.** Listening to body sounds is an **auscultatory method** of measurement. These sounds are related to the equalization of pressures between the arm blood vessels and the pressure in the heart. This technique has been in use for many, many years.

Sphygmomanometers can use air or mercury (Hg) within the gauge to determine the pressure. The pressures are reported in millimeters of mercury (mm Hg), even though most hospitals have disallowed the use of mercury gauges.

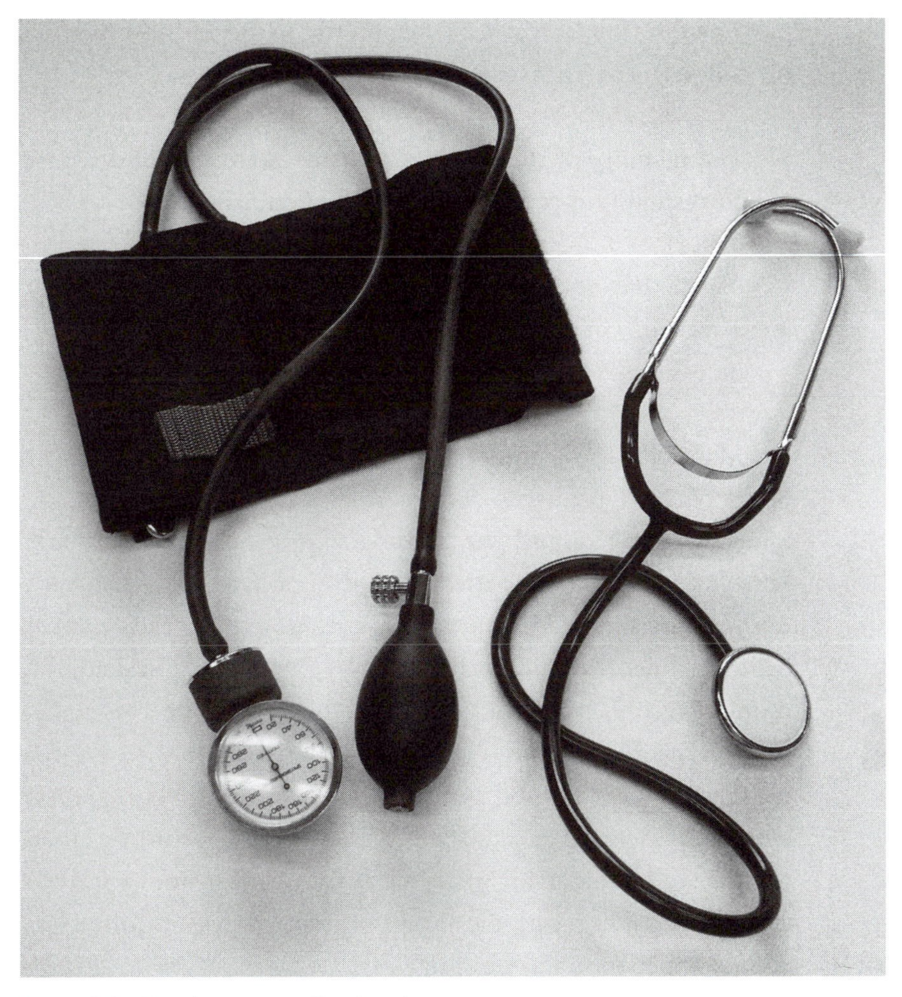

Figure 7.1. Blood pressure cuff and stethoscope.

Systolic pressure is the highest pressure in the heart and is measured when the heart is contracting. **Diastolic pressure,** the lowest pressure in the heart, occurs when the heart is filling. Blood pressure typically is reported as a two-number ratio, systolic "over" diastolic pressure. Typical adult pressures are 120 mm Hg for systolic pressure and 80 mm Hg for the diastolic pressure.

There are *limitations and sources of error* in this indirect method of measurement. Measurements can be compromised by

- The inability of the observer to listen accurately and record corresponding pressures accurately (ambient noise can be a factor)
- A cuff that does not correctly match the size of the patient
- The health of the patient - given that good limb perfusion is necessary

Many clinical facilities use **automated** equipment to measure blood pressure primarily because nurses may not be the person taking vital signs, and the automated system is seen as more reliable. Many times, patients' vital signs are recorded by a medical assistant, like a nursing assistant, visiting each patient bed at defined time intervals. Figure 7.2 shows a physiological monitor (NIBP, temperature) on a rolling pole. The automated measurement technique is meant to allow people with limited training to record blood pressure. As the nursing shortage has persisted, assigning tasks such as measuring and recording blood pressure to less-skilled workers has been a strategy for continued quality of patient care. Using automated devices requires less clinical experience and training.

Some staff may call all NIBP monitors Dinamaps since that is a common brand. Many **Dinamaps** are mounted on rolling poles to be moved from patient bed to bed. In addition to blood pressure, most physiological monitors can assess other patient variables like temperature and heart rate. Many NIBP devices are designed to be moved from bed to bed by a nursing assistant to record the vital signs (not just blood pressure) of patients at regular intervals (often every four hours).

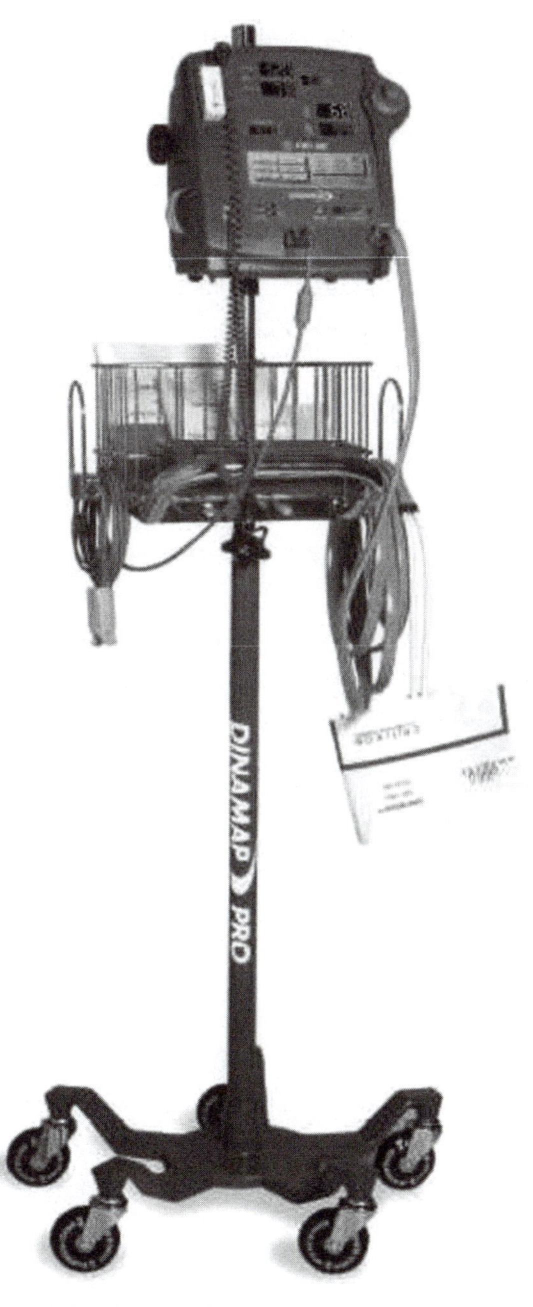

Figure 7.2. Dinamap physiological monitor.

Automated devices use a blood pressure cuff like that used for the manual method, but the cuff has no gauge and has a special connector to attach to the NIBP device. However, these devices detect changes in blood pressure using a method that does not listen for Korotkoff sounds (auscultatory method) but instead detects tiny pressure fluctuations from the arteries of a patient inside the blood pressure cuff. First the device inflates the cuff to a typical value (which may be related to the previous blood pressure measurement). The device deflates the cuff in small increments, matching these pressure oscillations to slowly deflate and determine the systolic and diastolic pressures. This process is called the **oscillometric method**.

Direct invasive arterial pressure measurement

Blood pressure can also be measured *directly* by placing a catheter into one if the patient's blood vessels. This may be done in the arm and used when constant monitoring is critical for patient treatment decisions (patients in intensive care, for example). This technique involves direct measurement of arterial pressure by placing a catheter (small tube) in an artery (often radial or femoral). The arterial catheter must be connected to a sterile, fluid-filled system that is connected to a sensor. A pressure sensor, such as a strain gauge or piezoelectric crystal, can measure the pressure in the fluid-filled tubing. The resulting waveform is often displayed on a monitor, as is shown in Figure 7.3. The bump in the pressure waveform is called the *dicrotic notch*. It is a pressure change related to the aortic valve closing. The timing of this notch can provide clinicians with additional patient heart performance information.

Mean arterial blood pressure (MAP) is used in the clinical setting to describe blood pressure as one number that represents the

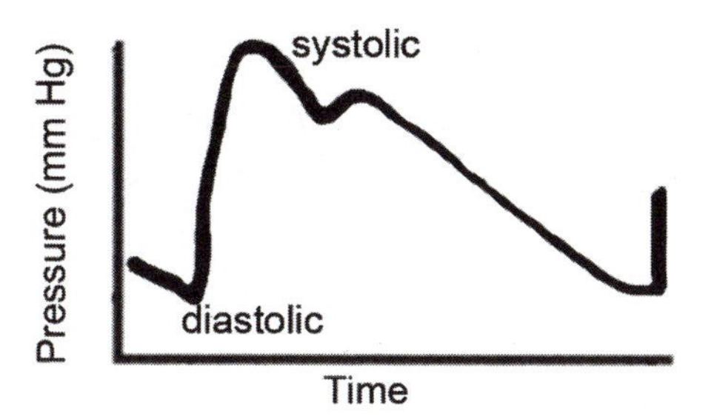

Figure 7.3. Typical blood pressure waveform.

average blood pressure in a patient. Typical values are approximately 100 mm Hg. Although the directly recorded blood pressure waveform can be used to calculate this number, MAP is not usually a mathematical average. There are several ways to calculate MAP, but, when monitoring a patient, it is generally done by the ECG monitor using the geometric mean of the pressure waveform.

Many different pressures within the heart can be measured directly using a **Swan-Ganz catheter.** To insert this device, a thin tube is threaded into the pulmonary artery and measurements of heart function are recorded when the catheter is in the proper location. The catheter can assist in determining the pressure inside the right side of the heart and in the pulmonary artery as well as in establishing the cardiac output. These devices may remain connected to the patient in an intensive care unit as a constant monitoring device for critically ill patients.

STUDY QUESTIONS

1. List and describe the advantages and disadvantages of NIBP as opposed to the auscultatory method of blood pressure measurement.

TABLE 7.1. *Blood pressure ranges*

Vital sign	Infant	Child	Adult
Blood pressure (mm Hg)	90/50	125/60	95/60 to 140/90

2. Create a memory aid to identify systolic and diastolic pressures. In typical blood pressure reporting, identify which pressure is reported first and which is reported second (as shown in Table 7.1).
3. Identify some of the advantages and disadvantages of continuous arterial blood pressure monitoring.
4. If mean arterial pressure was calculated using a strict mathematical average, what value would the MAP have?
5. List and describe the physiological measurements that can be determined using a Swan-Ganz catheter.

FOR FURTHER EXPLORATION

1. States have restricted or banned the use of mercury. Therefore, most blood pressure gauges do not contain mercury. Use the Internet to explore how this ban affected hospitals as blood pressure devices and thermometers were restricted. Describe the historical changes that have taken place. Reflect on the continued measurement of blood pressure in millimeters of mercury when mercury is a banned metal.
2. Use the Internet to record the medical definition of hypertension for adults. Most NIBP devices have limits of 300 mm Hg. Is it possible that a patient could have a blood pressure higher than this value? If so, under what circumstances? How could blood pressure be measured if it is greater than 300 mm Hg?

3. Blood pressure cuffs are sized to match the patient. Use the Internet to explore the various sizes of cuffs. List the various sizes that are commonly available. Describe the cuffs designed for pediatric patients. Describe what happens when an obese patient has his blood pressure taken with a cuff that is too small.
4. Blood pressure cuffs may be disposable or reusable. Discuss the contamination issues that surround reusable cuffs. Are they usually disinfected between patients? How is this commonly used device involved in hospital-acquired diseases? Use the Internet to support your answers.
5. Discuss the economic advantages of the use of all-in-one devices that can easily be used to take the vital signs of patients. Identify the amount of training that is likely needed to operate a typical device. How does this training compare to the training needed by a nurse? How does the hospital benefit when it uses all-in-one devices and low-paid staff to record vital signs? Are there disadvantages related to this approach?
6. Examine the GE Web site that explores Dinamap technology, reading the pdf file *The Dinamap Difference: A Guide to Our NIBP Technology* http://www.gehealthcare.com/usen/patient_mon_sys/docs/DINAMAPDifference.pdf. Interpret and summarize the information provided to explain in detail the process of determining blood pressure using an NIBP device.
7. Blood pressure is related to the volume of fluid in the body. Use the Internet to research and summarize the relationship between hypotension and critically injured patients. Identify and describe the situations that can cause hypotension. Identify and describe treatments that will improve blood pressure.

8. Blood is pumped/pushed through the arteries. Blood is returned to the heart in a passive system that uses the veins. Use the Internet to research and summarize the role of venous valves in the return of blood to the heart. Identify conditions where this return can be hindered.

8

Respiration and respiratory therapy

LEARNING OBJECTIVES

1. list and describe the components of inspired and expired air
2. list and describe the important lung volumes
3. describe a spirometer
4. describe impedance plethysmography
5. describe apnea
6. describe capnography
7. list and describe the functions and settings of a ventilator
8. describe and characterize high-frequency ventilation
9. describe the method of connection between patients and the ventilator
10. describe nebulizer, oxygen tent, and humidifier

Introduction

The focus of this chapter is *external human respiration*. Moving air in and out of the lungs is external respiration and the focus of technological interventions. The exchange of gases in the alveoli is part of internal respiration, but it is not essential to this chapter. Here are some important facts related to respiration:

Make-up of air: 79% nitrogen, 20.96% oxygen, and 0.04% carbon dioxide

Make-up of expired air: 79% nitrogen (unaffected by respiration), 17% oxygen, and 4% carbon dioxide (human's waste product)

It is important to know that the exchange of gas in the lungs occurs in the alveoli.

Respiration measurements

Volumes of gases that fill the lungs during different moments in respiration are important to BMETs since many machines try to fill those volumes. Amounts vary by the age, height, gender, and physical condition of a patient.

Tidal volume: Breathing volume is about 500 milliliters (mL) for an average adult male. This is the main value of importance. Figure 8.1 shows a sine wave that illustrates the action of breathing in and out. The tidal volume is moved in and out during each normal breath. When technology is used to support breathing, tidal volume is the amount of air that the machine must deliver.

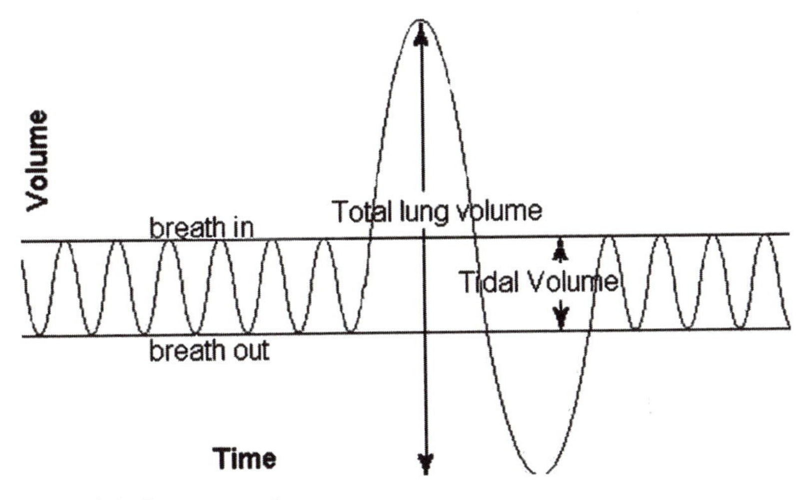

Figure 8.1. Respiratory volumes.

When measuring breathing activity, most volumes are measured using *flow rates* rather than actually measuring the volume. However, **spirometers** are used to measure actual patient *lung volumes* by respiratory therapists, often at the patient bedside. More complex spirometers are electronically controlled and provide a great deal of information about the performance of the lungs of the patient. However, most automated measurements (in ventilators, for example) are done using flow rates.

The typical adult respiration rate is about 12 breaths per minute (see Table 8.1). **Hyperventilation** is breathing at a rate that is faster than normal. **Hypoventilation** is breathing at a rate that is slower than normal.

One important feature that is monitored is the *effectiveness of respiration* – a patient may be breathing (air moving in and out) but it is more important to know whether this activity is resulting in appropriate oxygenation. Lab tests (often performed at the bedside or in the unit) on patient blood can give such information. See

TABLE 8.1. *Respiration rate ranges*

Vital sign	Infant	Child	Adult
Respiratory rate – number of breaths per minute (breathing) – values are listed in a range for normal	30–50	18–30	8–18

Chapter 13 for more information. Pulse oximetry is also used to simply evaluate and measure the effectiveness of respiration.

Measuring respiration rates is important. This is often done using transthoracic (across the chest) **impedance** measurements (called **impedance plethysmography**) since lung volume changes with air – which is a dielectric and changes the impedance measured between two points. To do this, a small signal is passed through the ECG electrodes, which are located on the chest of the patient. The resulting electrical change in impedance is measured. The changes in impedance (with changing lung volumes) can be represented graphically over time as a respiration waveform.

Apnea is a prolonged pause in breathing. It is a common problem in premature infants and can be life-threatening. Premature infants are often monitored for apnea for months after they are born. Frequently, pulse oximetry, which does not measure respiration rate, is used to detect periods of apnea. This technique is used because it is simple. The decrease in oxygen saturation (during periods without a breath) as well as detection of the pause in breathing rate can trigger an alarm. Interestingly, episodes of apnea can be stopped by shaking the patient or tapping the incubator.

Capnography is the measurement of carbon dioxide concentrations in expired air. This technique evaluates the changes in carbon dioxide values over time. This is often done during general anesthesia in surgery. Capnography can provide data to monitor the patient response to anesthetic as well as the mechanical ventilation techniques required to promote effective oxygenation. The Web site http://www.capnography.com provides a very detailed explanation with some excellent graphics.

Mechanical ventilation

Long ago in medical device history, breathing was assisted using chambers that enclosed the body to surround the chest with lower than atmospheric pressure (negative pressure) to make it easier to breathe (termed *negative extra-thoracic pressure*). These chambers were called **iron lungs**. Iron lungs brought on problems by adversely affecting blood flow in the patient. Patient care and movement were also tremendously hindered because the body was confined to a chamber.

Mechanical devices today work in the *opposite* way – air is pushed into the lungs (termed *positive intra-pulmonary pressure*). This does not mimic actual breathing. In normal breathing, air is sucked into the lungs by moving the diaphragm.

Humans have ventilated other humans for centuries. There is documented evidence from the 1700s of fireplace bellows being used to force air into the lungs. Ventilation bags, which are similar to fireplace bellows, allow one person to squeeze the device, forcing air into a patient's lungs. They are commonly used for brief periods when technology is not available or practical.

Ventilators, also called **vents** and **respirators**, are mechanical devices that can either breathe for a patient or assist a patient's breathing activities. A picture of one model is shown in

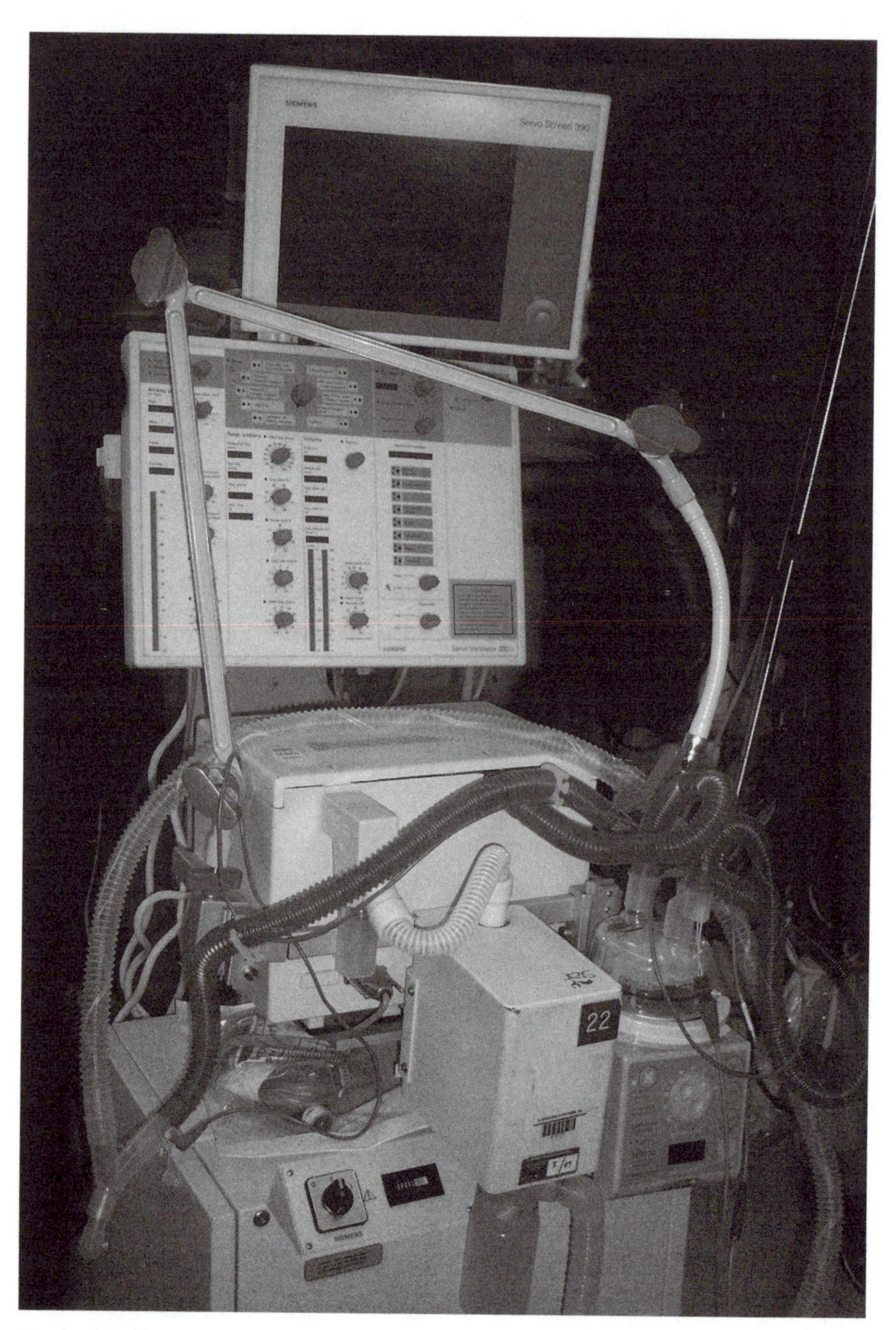

Figure 8.2. Ventilator.

Figure 8.2. They can be quite complex and are generally controlled by microprocessors. Breaths may be delivered at a preset rate (for patients whose respiratory systems are injured or intentionally paralyzed) if a patient goes too long without taking a breath or whenever the patient initiates a breath.

Patients with some respiratory function can have a role in ventilation with a ventilator. For example, the patient can initiate a breath and the machine will complete it. Careful monitoring of the patient's respiratory effort coordinated with device performance is vital.

Physicians order particular settings for their patients on ventilators, and respiratory therapists ensure that the equipment meets the doctors' orders. Often BMETs work hand in hand with the respiratory therapists since both spend a great deal of time with the equipment.

Some of the settings related to ventilators include:

- Respiratory rate – This setting determines how often the machine delivers breaths.
- CMV (continuous mandatory ventilation) – In this mode, ventilation occurs at regular intervals. This is often used when patients cannot initiate breaths on their own. Common rates are 10–15 breaths per minute for an adult. Note that this frequency in hertz (breaths per second) is less than 1 Hz.
- SIMV (synchronized intermittent mandatory ventilation) – This setting delivers occasional breaths produced by the ventilator, for patients who may be breathing on their own but not often enough to adequately oxygenate themselves. The breaths are timed to complement the breathing pattern of the patient.
- CPAP (continuous positive airway pressure) – This feature, which makes initiating a new breath easier, is often used as a

therapy that can be provided at home. This assistance is provided for patients who are breathing on their own but need some assistance.

- PEEP (positive end expiratory pressure) – For patients who are completely machine dependent, this setting is similar to CPAP.
- Oxygen concentrations – This measurement ranges from normal air percentages to 100% oxygen.
- Sensitivity – This setting determines how much effort is required for a patient to trigger a breath. It is useful when weaning patients from ventilator dependence.

The amount of air that is put into the lungs is important. However, most machines determine the amount of air not by measuring the volume but by measuring the flow rate and calculating the volume.

Ventilators are life-sustaining devices, and their reliable and consistent performance must be assured. Alarms and backup systems are integral to ventilation design to warn of unintended behavior or device failure. Ventilators can often function briefly on battery power. In addition, they are often connected to outlets, which ensures emergency power in the event of an outage.

High-frequency ventilation

High-frequency ventilation (HFV) was introduced in the early 1990s for use in the neonatal intensive care unit. There are two main types: the high-frequency **jet ventilator** and the high-frequency **oscillator ventilator** (HFOV). The goal of the HFV is to improve alveolar stability, oxygenation, and ventilation with decreased lung damage related to pressure and volume. The lungs *stay inflated* while both volume and pressure changes associated with continuous forced opening and passive closing of alveoli are avoided. With the oscillator ventilator, breaths are delivered by a vibrating diaphragm that provides for both a positive inspiration and active exhalation.

The settings for the HFOV differ from the conventional ventilator. Settings are:

- MAP (mean airway pressure) directly affects oxygenation by improving lung volume.
- Delta pressure (Delta-P) is the oscillatory amplitude, and, as the difference between the peak and trough pressures, it directly determines tidal volume. Delta pressure is used to control the carbon dioxide (CO_2) - increasing the Delta-P will cause the CO_2 to drop; decreasing the Delta-P will cause the CO_2 to rise.
- Hertz is the expression of breath rates: Hz is commonly set at 10–15, which is 600–900 breaths/minute. Adjusting the hertz will also change the tidal volumes.

Caution: This is a relatively new technique, and some older texts and Web sites may define high-frequency ventilation as any ventilation greater than hyper-respiratory rates, say 60 breaths per minute. This definition is *not* the same as HFV.

Ventilators are most commonly connected to the patient via an **endotracheal tube** through the mouth. The tube is curved to match human anatomy. Many endotracheal tubes contain an inflatable cuff at the end to make a seal against the sides of the trachea, as shown in Figure 8.3. The cuff is inflated with air from a syringe. A **laryngoscope** is used to guide the insertion of the endotracheal tube. These devices are very common in hospitals

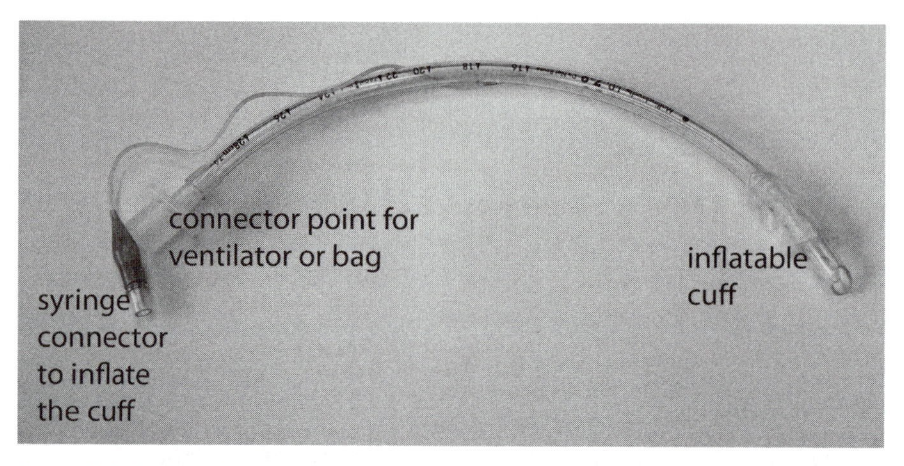

Figure 8.3. Endotracheal tube.

and can be lighted to guide the insertion of the endotracheal tube. A second type of ventilator connection, which goes through the neck, uses a **tracheostomy tube** (sometimes called a *trach*). The tracheostomy is typically used for patients who will be dependent on a ventilator for a long time.

Other respiratory equipment

The respiratory therapy department uses many devices that are not ventilators. These include oxygen tents and masks, humidifiers, and nebulizers.

- **Oxygen tent** – This device is usually made of plastic and is placed over a bed in order to maintain an oxygen-rich environment. Temperature and humidity may also be controlled in the tent environment. Oxygen tents are often used in pediatric units so that patients may move around.

- **Humidifier** – This device increases the humidity of the air and can be used in many situations. Humidifiers may be part of ventilators and incubators, among other devices.
- **Nebulizer** – This device is able to create a medication mist that can be added to air to be inhaled by a patient.

Extracorporeal membrane oxygenation (ECMO) is a method of oxygenating blood outside of the patient's body. Similar to a heart-lung bypass, which is done during open heart surgery, this device consists of a pump to move the blood, warmers and filters, and filters for the blood. A newer term for this device is extracorporeal life support (ECLS). This is a support method to help the lungs of neonates mature or heal. It is often done as a last resort when other mechanical techniques have failed. A connection is made directly from the patient's heart to the ECMO machine. Although more commonly used for infant patients, some children and adults are connected to ECMO devices.

STUDY QUESTIONS

1. Using Figure 8.1, compare the total volume of air in the lungs and the relative amount of air that is moved during normal respiration.
2. What is the main component of the air humans breathe? Is it a gas that is used by the body during the respiratory process?
3. Describe why it is much more informative to know about the quality of respiratory activity than the respiratory rate.
4. List and describe some of the technologies and tests used to monitor patients who have breathing support.

5. Describe how pulse oximetry can be used to monitor newborn babies for apnea.
6. Describe how CPAP can assist patients in the initiation of a new breath.
7. Describe the major differences between common ventilators and high-frequency ventilators.

FOR FURTHER EXPLORATION

1. Use the Internet to explore and document the volume of lungs in various patients including premature infants, toddlers, and athletes. Create a table showing the various lung volumes for different types of patients. Identify and compare the smallest and largest values. How do these values compare to the assumed average value of 500 mL?
2. Many people have CPAP machines at home. Use the Internet to explore and document how these devices work. Identify the parts of these systems. Describe how they benefit the user.
3. Explore and describe the historical use of the iron lung. This was desperately needed for polio patients. Research and document the relationship between polio and the iron lung. How did it assist the breathing of patients? How effective was it? Identify its side effects. Briefly explain the dramatic improvement in breathing support that was experienced with the invention of the ventilator.
4. Calculate the breathing rate of a typical ventilator in hertz. Compare the breathing rates (in hertz) for a typical adult ventilator with that of high-frequency ventilation. Discuss this significant magnitude of difference.

5. Use the Internet to explore, research, and identify the major components of an ECMO system. What are the primary functions of most ECMO systems? How do these functions differ from those of cardiopulmonary bypass?
6. Describe how ventilators can affect a patient's speech. What techniques can be used to allow patients to speak?
7. Breathing circuits are made up of tubing that connects patients to ventilators and anesthesia delivery systems. What are the components of breathing circuits?
8. Ventilators shown in the media often show a piston (a black accordion device) moving up and down (it makes a very nice visual representation of mechanical support). Often, an easily identifiable noise (cyclical, piston sound) can be heard in hospital scenes even when no ventilator is being used. These volume ventilators are generally no longer in use. Use the Internet to find a photo of a Puritan Bennett MA1. This device used to be state of the art and was extremely common. Describe the features of this device. Search the Internet for the Puritan Bennett 840, the company's latest ventilator. Compare the two devices.
9. To understand breath delivery by ventilators, visit the Puritan Bennett Web site http://www.puritanbennett.com/_Catalog/PDF/Product/ABCsSmarte%20BreathDeliveryClinicalBrochure.pdf, and look for the pdf file *Clinical Brochure: ABCs of Smarter Breath Delivery*. Summarize the information in this brochure.

9

The brain and its activity

LEARNING OBJECTIVES

1. list and describe basic neuroanatomy and physiology terms
2. list and describe EEG characteristics (four waves)
3. describe depth of anesthesia monitoring
4. describe ICP monitoring
5. describe a CSF shunt and identify its functions

Introduction

Measuring the activity and function of the brain is more complex than measuring the electrical activity of the heart. Signals from the brain are more random and have only recently been translated into specific applications. Even though brain signals can be extremely useful in patient assessment, in general, measuring the function of the brain is far less common than other technology-facilitated patient care tools.

Review of neuroanatomy and physiology

Here is a brief review of basic neuroanatomy and physiology terminology.

- Neuron – This is the most basic cell associated with the brain.
- Brainstem – This portion of the brain controls life-sustaining functions.
- Cerebellum and cerebrum – These major parts of the brain control various body functions and memory.
- Ventricles of the brain – These spaces within the brain are important because implants and monitors often involve these areas.
- Spinal cord sections – The spinal cord includes the cervical, thoracic, lumbar, and sacral regions.
- Cerebral spinal fluid (CSF) – This fluid is very important because it cushions the brain and can be monitored to assess brain health.

Electrical activity in the brain

Electroencephalography (or electroencephalogram)

The electrical signals from the brain can be measured using many electrodes on the scalp. EEG signals created inside the brain are complex and relatively random. Measured at the scalp, EEG amplitudes are very small, in the range of microvolts. Specifically, EEGs are used to:

- Evaluate brain lesions
- Identify epilepsy
- Evaluate mental disorders
- Assess sleep patterns
- Evaluate brain responses to stimuli

Patterns and information are more difficult to obtain from an EEG than from an ECG. Nevertheless, tests are often performed that evaluate the EEG waveform in response to an input like light or sound. Since data can be gathered from many people, a typical EEG response can be predicted and compared to the response of a particular patient. Noise, motion artifacts, and ECG signal interference can present some major issues because EEG waveforms appear at the scalp in such low voltages.

Brain signals, which are grouped based on frequency, are divided into alpha, beta, theta, and delta waves. Different signals are prominent depending on the age and activity of the patient.

- Alpha - 7.5–13 Hz; present during rest and relaxation.
- Beta - 14 Hz and higher; present for patients who are alert and awake. These waves have the lowest amplitude.
- Theta - 3.5–7 Hz; normal for awake children.

- Delta - 3 Hz or below; common during sleep and in infants. These waves have the highest amplitude.

Evoked potential testing

When an EEG is being evaluated, a common test is to measure the electrical activity of the brain in response to a stimuli. Some examples of stimuli include sound or a flash of light. When this test is performed, the EEG will be examined for the time the brain takes to respond and the location of the response.

Depth of anesthesia monitoring

The EEG can be used to determine the level of sedation of a patient during anesthesia. This noninvasive method analyzes the EEG using mathematical algorithms. The analysis technique was created using data from many test subjects who underwent anesthesia. The mathematical algorithm used to analyze the EEG from three EEG leads on the forehead of the patient varies by companies. Figure 9.1 illustrates the placement of leads on the forehead. A popular method of analysis is **BIS**, which stands for Bispectral Index. The monitoring produces a numerical score of 0 (dead) to 100 (fully awake). This monitoring allows careful control of sedation and more precise delivery of medication. The BIS monitoring technology is used by multiple manufacturers.

Intracranial pressure (ICP) monitoring

The ICP is essential in the treatment of brain-injured patients. Pressure can be monitored under the skull or in the ventricles. The pressure is normally about 1–15 mm Hg in adults (20 mm Hg is usually considered the limits of safe pressure) and is typically measured in the ventricles. ICP monitoring can be done in several ways.

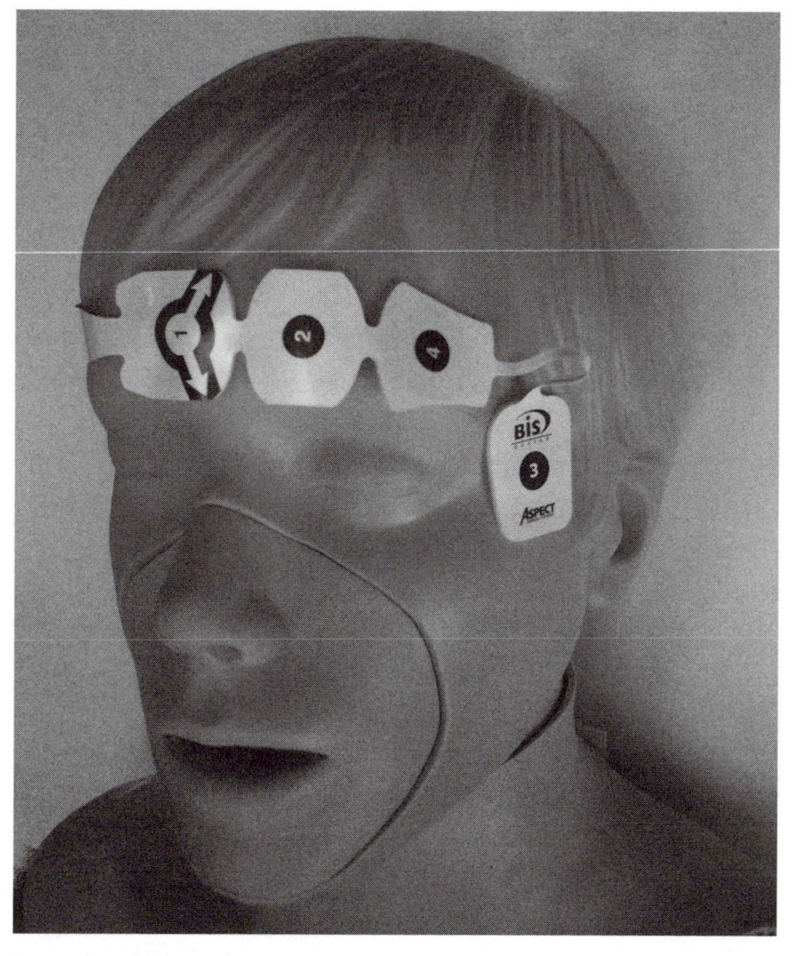

Figure 9.1. BIS leads on a patient.

- Subdural/subarachnoid pressure can be monitored by attaching a bolt to the skull. The data may not be very reliable, but the ventricles are not entered so infection problems are avoided.
- An intraventricular catheter can be inserted through the frontal lobe of the brain and into the lateral ventricle. The catheter can be used to monitor the ICP as well as to drain CSF.

- A fiber-optic ICP sensor located at the end of a catheter can be placed in the patient's parenchyma (Camino ICP monitoring is the most common method).

Calibration/zeroing is required for these transducers.

Brain perfusion

Understanding and evaluating the blood flow to the brain is a useful diagnostic tool in the evaluation of patient condition. Cerebral perfusion pressure (CPP) is the intracranial pressure subtracted from the mean arterial blood pressure. However, as a diagnostic tool, **cerebral blood flow** (CBF) is a better indicator of neurological health. Currently, this is not easy to measure. Some emerging techniques include xenon-enhanced computer tomography (XeCT), positron emission tomography (PET) scans, and single photoemission computer tomography (SPECT). Thermal dye-dilution can be used but is not used commonly at the bedside.

EEG monitoring

This type of EEG testing can be done at the patient's bedside to diagnose conditions related to coma and seizures and to evaluate recovery. There are quite a few problems with this type of testing because the environment in the intensive care unit is not controlled as well as that in the EEG laboratory. Some sources of difficulties include:

- 60-Hz noise
- Ventilator, IV pump, and vibrating mattress artifacts
- Movement and contact by nurses and therapists
- Patient movement
- Sweat and muscle artifacts
- Wounds or burns on the patient's scalp that limit access

Sleep studies

EEG monitoring is used extensively to assist in the diagnosis and treatment of sleep disorders and sleep-related respiratory difficulties. Patients are connected to an EEG recorder and other monitoring devices and asked to spend a night in the hospital for observation. Sleep study rooms are monitored by technicians who can record data and make adjustments to patient equipment such as CPAP settings. Figure 9.2 shows a patient ready for observation. The data from the leads connected to the patient will be processed on an EEG recorder. Figure 9.3 shows one type of recorder. Figure 9.4 shows a homey sleep lab room where observation can take place.

Cerebral spinal fluid shunts

Cerebral spinal fluid shunts are used to transfer excess fluid from the ventricles into the abdomen (or other areas). When the brain has structural anomalies, CSF shunts are implanted to avert buildup of cerebral spinal fluid. These devices are essentially plastic tubing that drains fluid from the brain (using gravity) into other parts of the body where the CSF can be absorbed.

STUDY QUESTIONS

1. Compare the amplitude of an ECG and the amplitude of an EEG. Identify difficulties that may be encountered because of the small EEG amplitude.
2. Evoked potential testing can be used to test the hearing of infants. Describe how using sound and evaluation of brain waves could be used to evaluate hearing in nonverbal infants.

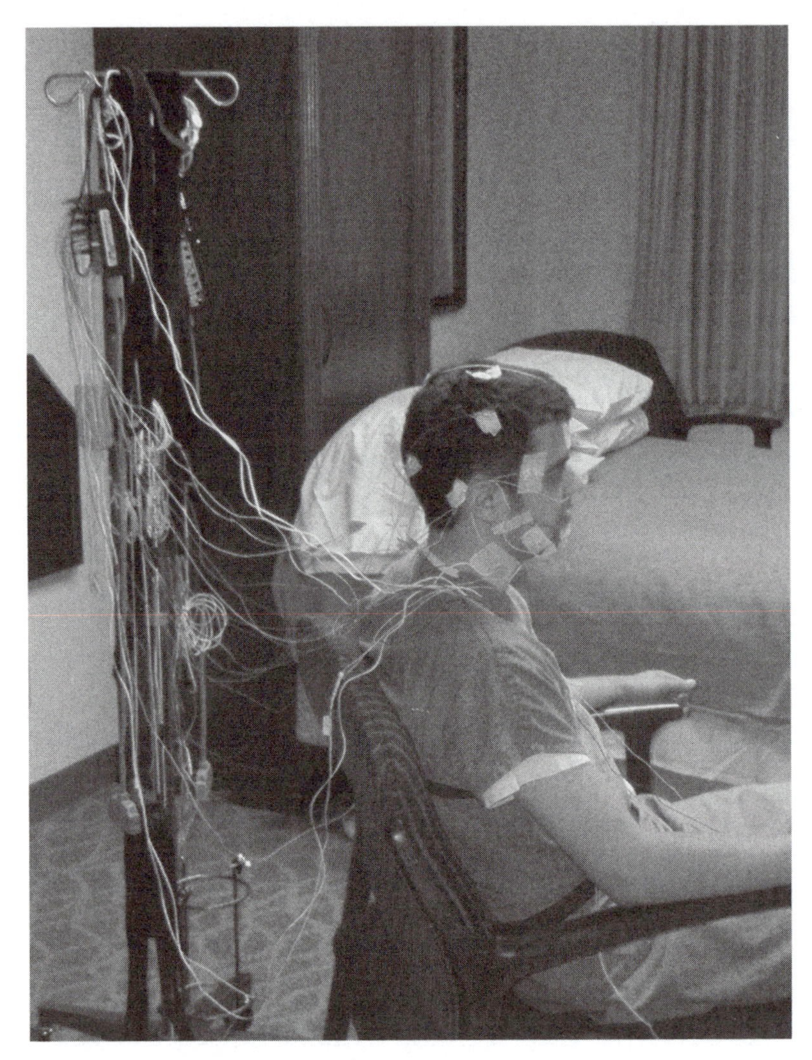

Figure 9.2. Patient connected to EEG leads.

3. Describe the advantages of using depth-of-anesthesia monitoring using EEGs.
4. The skull is rigid so brain swelling is dangerous because, unlike swelling in other parts of the body, the organ has limited space.

Figure 9.3. EEG recorder.

Describe pressure monitoring and how this can be beneficial for patients.

5. Assessing the brain activity of unconscious patients is helpful. Discuss some of the difficulties in measuring the electrical activity of the brain.

Figure 9.4. Sleep laboratory patient care/observation room.

FOR FURTHER EXPLORATION

1. Use the Internet to obtain a drawing of the vertebral column and spinal cord. Identify the cervical, thoracic, lumbar, and sacral regions. Sketch this and be able to identify the regions of the body that the various spinal nerves control. Which is closer to the brain, T3 or L1? Spinal cord injury can result in paralysis. Research and document the types of spinal cord injuries (and at what level) that result in quadriplegia.
2. Epidural anesthesia, commonly used in the surgical delivery of infants, is a technique of applying anesthetic to an area around the spinal cord. Research and document the process used in this technique, including the location of medication delivery along the cord.

3. Create a table that identifies the four types of EEG waves and their amplitude and frequency characteristics. Include, within this table, details related to the conditions under which the particular waveform is usually present.
4. Use the Internet to research and document the portions of the brain that generate different EEG waves.
5. Investigate and document four pathological conditions that can be reflected in abnormal EEGs. Include epilepsy and dementia and describe the symptoms of the condition, how the EEG is altered, and the possible treatment considerations.
6. Research and document the discovery of EEG waves. Include which wave was identified first and how this was accomplished. Cite your references.
7. Investigate and identify the equipment used to deliver evoked potentials. What types of stimuli are used? How is this information useful?
8. Some neurological research involves the measurement of electrical activity directly on the surface of or inside the brain. Research, document, and describe this technique, which is typically called intracranial EEG. Identify the purpose of this process.
9. The muscles of the body generate voltage, and this can interfere with the measurements of EEG waveforms. Movement of the eyelids or tongue can cause interference. Research and summarize the methods to obtain EEGs from patients. Specifically discuss what techniques are used to eliminate the effect of muscle voltages as interference.
10. Before the use of depth-of-anesthesia monitoring (which provides quantitative data), anesthesiologists relied on qualitative assessment methods. Research and document these qualitative

techniques. How is a numerical system an improvement in anesthesia delivery?

11. Brain Computer Interface (BCI) has gained attention in the press recently. Explore this technique to circumvent damaged spinal cords or sensory organs. Describe the technical requirements and provide some examples of BCI applications.

10

The intensive care unit

LEARNING OBJECTIVES

1. identify and describe the characteristics of an ICU and an ICU patient
2. identify and describe the types of monitoring that occurs in ICUs
3. identify and describe the major types of ICUs and the support technology used within them
4. define and describe medical telemetry and WMTS

Introduction

Patients who are very ill or in serious condition will usually be found in the intensive care unit (ICU). Although these patient care areas may be called various names, they have some similar qualities. Generally, intensive care units have a very large amount of technology involved in patient monitoring and treatment. Technical staff who work in this area need to understand the characteristics and special needs of ICUs.

Many hospitals in the United States have their own way of organizing and providing intensive care. For example, some hospitals do not separate surgical and medical patients. Some hospitals include transplant patients in their medical units, whereas others send transplant patients to the surgical floor, and still others have stand-alone transplant intensive care units. It is virtually impossible to address all the variations in this chapter. However, the issues important to intensive-care patients and staff who care for them transcend the hospital-defined categories.

Characteristics of intensive care units

Some smaller or rural hospitals may not provide a high level of intensive care; consequently, patients who need intensive care are transferred to other hospitals. Here are the **main characteristics** of ICUs:

- Patients are very ill.
- Patients have a high mortality rate.
- Patients may not be conscious or mobile.

- Most patients are restricted to bed.
- There may be limited day/night cycles.
- Physiological monitoring is constant.
- Each staff member is assigned to only a few patients.
- Units may be noisy.
- Patient may be in an ICU for weeks or months.
- There are restrictions about who can enter a patient room and when.
- Patient must be monitored very closely.
- Tubes attached to patients include feeding tubes, endotracheal tubes, IVs, and urinary catheters.
- Ventilators – patients may be dependent on technology to get air into and out of their lungs

Within the ICU, **monitoring** often consists of:

- The heart's electrical signals (often with either a three- or five-lead ECG)
- Noninvasive blood pressure
- Invasive blood pressure that uses an arterial line (a catheter that enters an artery)
- Blood gases – from a blood sample to evaluate the amount of dissolved gases it contains (often done at the bedside)
- Respiration rate
- Pulse oximetry – a noninvasive method of determining the amount of oxygen carried in the blood
- Temperature
- Chemistry of the blood – from blood samples to evaluate characteristics such as pH (often done at the bedside)

Patients generate a great deal of data that is often networked and stored for analysis later. Physicians can use trends in patient data over time to better diagnose and treat patients.

Types of intensive care units

Special care units where the patients and the staff differ from the rest of the hospital go by many names. A list of these special care units follows; however, be aware that sometimes individual hospitals will define the same letters differently.

Cardiac care unit

Patients in the **cardiac care unit (CCU or CICU)** may be recovering from a heart attack or bypass surgery. They are typically older and may have a high level of mortality. Additionally, they may have many characteristics of the elderly, including slower healing, general physical limitations, and less mental acuity. Cardiac patients may need to be resuscitated far more often than the general patient population. **Resuscitation** usually involves defibrillation. **Code carts**, which contain defibrillators and medications to treat nonperforming or underperforming hearts, are located in almost every room.

Equipment frequently found in the CCU includes balloon pumps and external assist devices for the heart. Balloon pumps are inserted into the descending aorta to help the heart function, which allows a patient's heart to rest and heal. The balloon inflates during diastole to increase pressure. The balloon deflates just before systole, creating a void, thereby lowering pressure and making it easier for the heart to pump. Cardiac assist devices, such as left ventricular assist devices or ventricular assist devices are essentially external pumps that can help the heart heal or support a patient until a transplant is found.

Pacemakers may also be used to support patients in the CCU. They may be temporary and external. Pacemakers are described in Chapter 6.

Neonatal intensive care unit

Patients in **neonatal intensive care units** (**NICU**, often pronounced as one word: *nick-you*) are typically between 23 and 40 weeks of gestation (normal pregnancy is 40 weeks) and can weigh from around a pound to more than 10 pounds. The small size and critical condition of these infants can be shocking when a BMET first visits the unit. Patients may have very red and translucent skin, need massive support for bodily functions, and be extremely small. Babies are divided into two large groups: those who are born before term and have complications that stem from their early birth and those who are born at term. Patients in the second group either have disorders or complications developed during the pregnancy or complications that were a result of the birth process.

Neonates need a very controlled environment in order to allow organs and systems to grow and mature. There are generally two types of environments used, incubators and warmers.

- **Incubators** (also called isolettes) are small beds completely enclosed by Plexiglas. Doors in the incubator allow clinicians access to the patients while minimizing heat loss. The air inside is usually warmed and humidified; it may also be oxygenated. An incubator is shown in Figure 10.1.
- **Warmers** are open beds that have a radiant heater and bright lights above the uncovered patient bed. A sensor

Figure 10.1. Incubator.

placed on the baby controls the temperature provided by the heater above. The patient bed is usually x-ray transparent so that x-rays can be taken without moving the patient. An example of a warmer is shown in Figure 10.2.

Some specific therapies that are most commonly used on NICU patients include ECMO (see Chapter 8) and treatment for jaundice. Infants may have hyperbilirubinemia, which is *jaundice*. The treatment for this is called *phototherapy*, which involves exposing the patient to a specific wavelength of light (in the blue region). Phototherapy treatment usually uses the wavelength of light that can travel through the skin and destroy the hazardous compounds in the blood. Before the introduction of fiber-optic

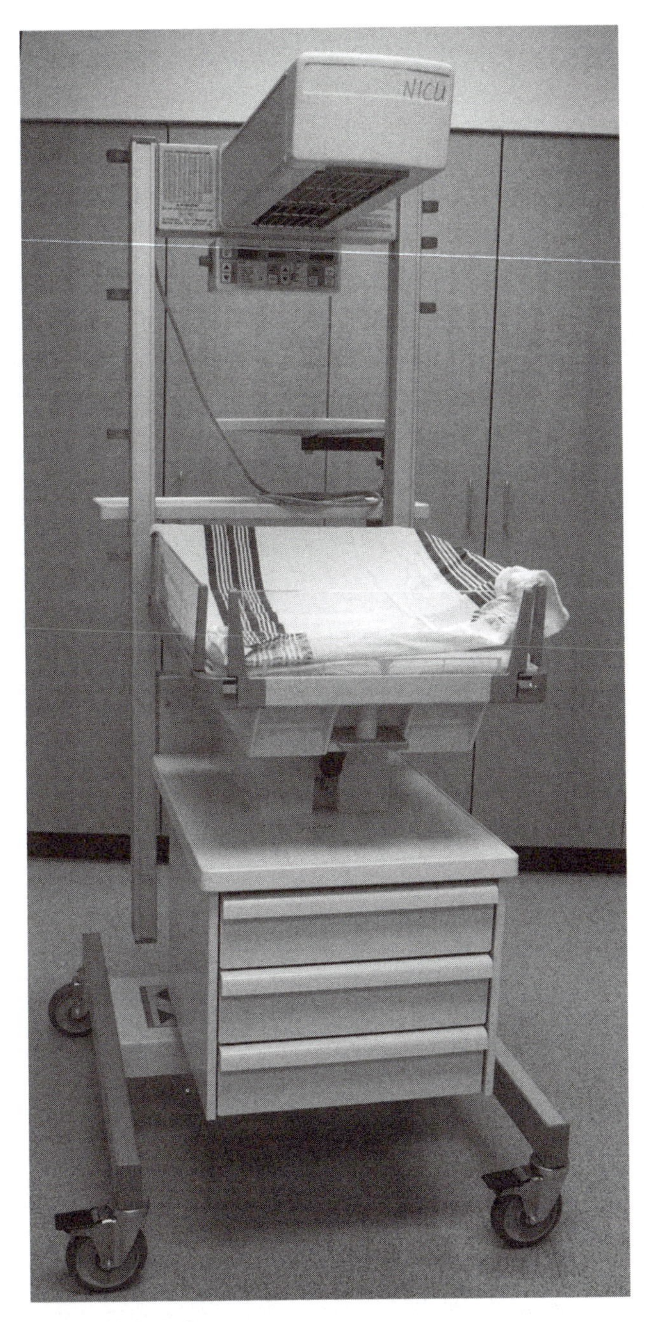

Figure 10.2. Infant warmer.

blankets, it was common to see banks of blue light bulbs shining down on the patients. The blue lights were, however, difficult for staff (they can cause nausea) and could mask the condition of patients. The use of phototherapy blankets is less intrusive to the NICU. The measurement of light intensity at the patient position is a critical step in the assurance of proper dosing of light. Light meters that measure in the correct wavelength (425–475 nm) can assess the light intensity. Fluorescent bulbs do decay in their light output, and BMETs may test at regular intervals to determine replacement frequency.

Pediatric intensive care unit

Pediatric intensive care units (PICU, often pronounced as one word: *pick-you*) are specifically for children. Children are not just smaller adults. Not only are their organ systems growing and changing, but the patients may be unable to be cooperative or answer questions. PICU patients are generally aged from newborn through adolescence (although some hospitals use age 1 as the youngest patients in pediatrics). Generally, the infants in a PICU may have been discharged at birth and then returned to the hospital. Patients who are in pediatric intensive care units may be recovering from surgery, have a serious illness, or have experienced trauma (such as drowning or other accidents). Many patients present with complex medical issues that involve several organ systems.

Medical intensive care unit

Patients in the **medical intensive care unit** (MICU) are seriously ill for medical reasons. Some examples include asthma, pneumonia, tuberculosis, and AIDS, but MICU patients also may have experienced trauma. The trauma a patient might have experienced varies greatly depending on the institution and its surrounding community. For example, gunshot wounds and drug overdoses may be common in an urban setting, whereas snake bites and scorpion stings might be common in another. Patients in this unit have a wide variety of complex health issues related to environment and disease, but they generally exclude specific and isolated issues with the cardiac and nervous system.

Surgical intensive care unit

Generally, patients in the **surgical intensive care unit (SICU)** are post-operative. Some may be critically ill as a result of surgery; others may be recovering from a surgical procedure and require the monitoring the unit provides. Hospitals may include neurosurgical or cardiovascular surgical patients in this group, and transplant patients may be included as well. Examples of surgical procedures include amputations, gastrointestinal procedures, peripheral vascular surgery, orthopedic procedures, and urologic surgery.

Many times, recovery from surgery is complex due to a patient's preexisting conditions such as diabetes or heart disease. Wound healing and closure is vital for patient recovery. Unfortunately, promoting healthy healing is not always easy, especially for patients with multiple medical issues.

Other types of ICUs

Burn patients may be cared for in a burn ICU. Burn patients may be treated with hyperbaric therapy. This approach to wound healing uses chambers that enclose the patient to increase the amount of oxygen carried in the blood. This is accomplished by increasing the pressure inside the chamber.

Patients who have neurological issues may be treated in a neuro ICU. Much of the equipment in this unit is described in Chapter 9, which explores the brain.

Support common for many patients

Deep Vein Thrombosis Prevention: A patient's inability to move may contribute to the development of deep vein thrombosis (DVT). DVTs may become life threatening if the thrombus travels to the lungs, causing pulmonary tissue death. The use of *sequential compression devices* is widely accepted in the ICU patient population to assist in the prevention of DVTs. These devices compress the feet or legs in a rhythmic way.

Nutrition: Patients who are critically ill need to consume calories, but they may be unable to do so because they are unconscious, injured, or lack physical strength. There are two sources for providing calories. **Enteral nutrition** provides calories through a tube into either the stomach or the intestine. A naso-gastric tube is often used to deliver the liquid nourishment. A tube passing through the nose may be a temporary feeding pathway; however, long-term support can be provided through a port in the abdominal skin, directly connected to the digestive tract. Nutrition may also be delivered via total parenteral nutrition (TPN). **Parenteral nutrition (hyper-alimentation)** bypasses the digestive system. This delivers a glucose and fat emulsion with other nutrients directly into the vascular system. An IV pump (feeding pump) is usually used to deliver this fluid.

Renal Support: Kidney dialysis replaces the functions of a normal kidney. Both water and waste products normally filtered

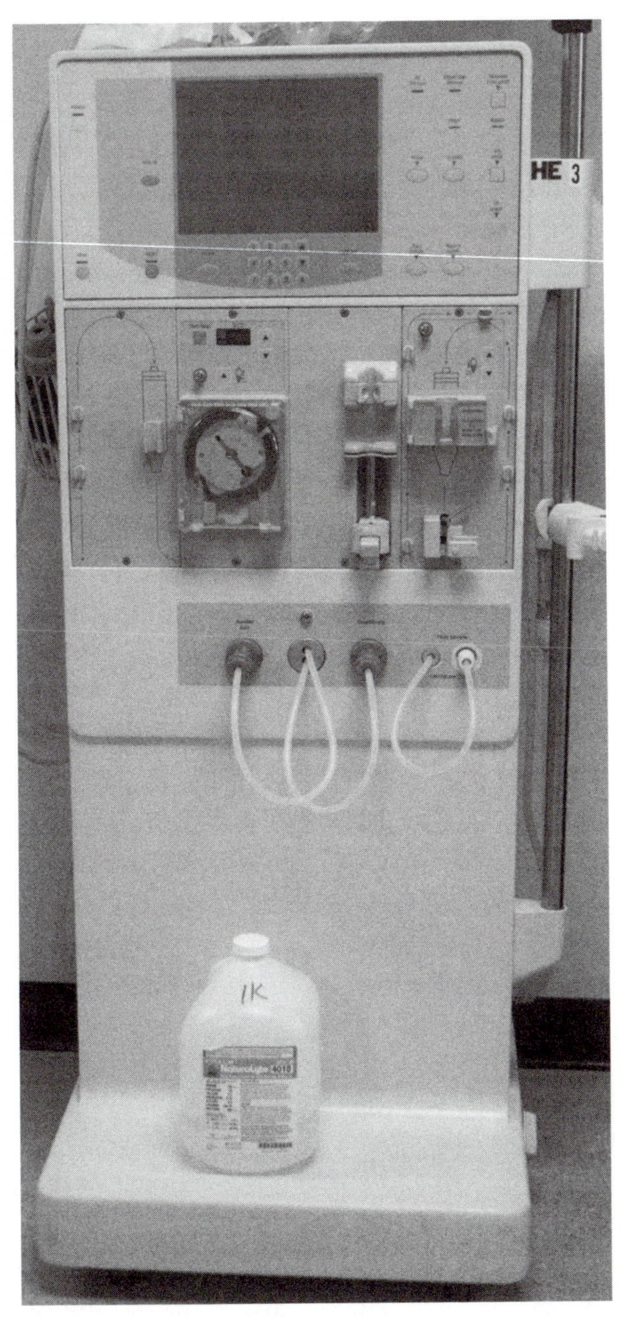

Figure 10.3. Dialysis machine.

by the kidney can be removed from the patient's body. Kidney dialysis machines filter blood using a semipermeable membrane. (See Figure 10.3.)

Patient Care Beds: In the past, patient care beds were devices with a mattress and a hand crank to raise/lower the head or foot. Now, beds have microprocessor controls that can monitor patient movement and inflate and deflate the mattress to alleviate pressure points. Hospital beds may have tiny holes in the covering to circulate air around patients. In short, patient beds are quite complex.

Telemetry

Patients who may be improving but still need constant monitoring may be allowed to get out of bed and walk around using telemetry. The physiological signals of the patient are transmitted over radio frequency waves (this technology was initially created for astronauts). The patient wears a transmitter, and the signals are sent to a central monitoring station (often at the main desk of a patient care area). The frequency that carries the signals has been challenging in a world where specific bands are set aside for applications by the Federal Communications Commission (FCC). Recently, a Wireless Medical Telemetry Service (WMTS) frequency band was designated. The FCC has designated the American Society for Healthcare Engineering as the medical telemetry frequency coordinator. Visit http://www.ashe.org and view its telemetry links. Originally, hospitals used 450–470 megahertz (MHz) until a test of Digital TV (DTV) stations in Dallas, Texas, in 1998. There was interference, and this began a discussion of setting aside a specific spectrum for medical telemetry. Now specific bands have been dedicated to WMTS 608–614 MHz, 1395–1400 MHz, and 1427–1432 MHz to limit interference.

STUDY QUESTIONS

1. Create a table where the rows are the different types of intensive care units and the columns are patient characteristics and specialized equipment. Briefly summarize information specific to intensive care units to complete the table.
2. Make a list of typical equipment found in a typical, general ICU.
3. Identify and describe the technologies that can be used to assist neonates with temperature regulation.
4. Describe and compare the pressure inside the chamber of an iron lung (Chapter 8) and the pressure inside a chamber designed to promote wound healing.
5. Compare and contrast the two methods of nutritional support.
6. Describe the benefits of telemetry to the patient.
7. Describe the types of physiological information that might be gathered over time and used by medical staff to support patient care decisions.

FOR FURTHER EXPLORATION

1. The Society of Critical Care Medicine offers an excellent Web site that provides a virtual tour of an ICU room. The site is http://www.icu-usa.com/tour/icu_room_tour.html. Visit this Web site to see the components of an ICU room. Describe and summarize your tour.

2. Continuous cardiac output (CCO) is a common CCU patient technology. Research and describe this technique, which uses a Swan-Ganz catheter. Describe how this technique is able to quantify the performance of the heart. Document typical patient measurements obtained using this technique.
3. For neonates, gestational age identifies how long the infant spent *in utero*. Gestational age may be a predictor of health and technological support needs. Use the Internet to research and document the relationship between viability and gestational age. What is the current minimum pregnancy duration that results in a live infant birth? Document how it has changed over time.
4. Birth weight can be a predictor of infant viability. Research and document the weight associated with very-low-birth-weight (VLBW) babies. Use the Internet to explore, research, and document the specific technologies involved in the care of very-low-birth-weight babies. Focus on equipment and devices (avoiding lengthy and complicated medical interventions).
5. The treatment for hyperbilirubinemia involves the exposure to the blue wavelengths of light. Use the Internet to research this medical condition and the technology used to treat it. Why do newborns have this condition? What does light shined through the skin accomplish? What wavelength is therapeutic? What technology is used to provide light therapy? Why do babies who are receiving this treatment have their eyes covered?
6. For children age 4 and younger, abuse and neglect are the leading causes of death. Pediatric units must have equipment to serve patients who have multiple system injuries (that is, brain injuries, respiratory problems, broken bones, and burns). Research, document, and describe the types of equipment required to support these patients. Discuss how these

multisystem injuries influence hospital care, equipment purchase decisions, and hospital spending priorities. Reflect briefly on the community profile that is served by a particular hospital and how the population of children might influence equipment purchasing decisions.

7. Hyperbaric oxygenation is used to increase wound healing, including burns. Use the Internet to research and document the many methods of exposing patients to high concentrations of oxygen to improve healing. Summarize the therapy methods, the equipment used, the benefits to patients, and the type of patients served.
8. Research and summarize the interference to medical telemetry caused by the DTV test in Dallas in 1998. Describe what happened. Discuss the changes that occurred after this problem. Define and describe WMTS. How will a dedicated frequency spectrum (WMTS) avoid these types of problems?

11

The operating room

LEARNING OBJECTIVES

1. list and describe the characteristics of surgical lights
2. list and describe the two functions of an anesthesiologist or CRNA (monitor patient and deliver medications)
3. list and describe the four functions of anesthesia
4. list and describe the staff in an OR
5. list and describe the items that must be worn in the OR
6. list and describe the four types of surgical procedures
7. list and describe the many surgical specialties
8. list and describe the two common methods of equipment sterilization
9. list and describe the stress related to the operating room environment
10. list and describe where patients go immediately after surgery
11. list and describe the function of the following pieces of OR equipment: depth-of-anesthesia monitors, operating microscopes, robots, smoke evacuators, patient warmers, blood and fluid warmers, operating room tables
12. list and describe laparoscopic surgery and the devices used in it

13 list and describe the functions and parts of an electrosurgical unit

14 define the term LASER and describe some common medical LASERs

Introduction

The operating room (OR) in a hospital is a unique environment that contains a great deal of equipment as well as many protocols and rules that impact a BMET. Many types of surgery are performed here. Operating rooms may be located in a hospital, in a free-standing surgery center, or in a doctor's office. BMETs must have a solid and accurate understanding of the operating room in order to perform their job effectively. A typical operating room is shown in Figure 11.1; it bears little resemblance to the operating rooms depicted in the media.

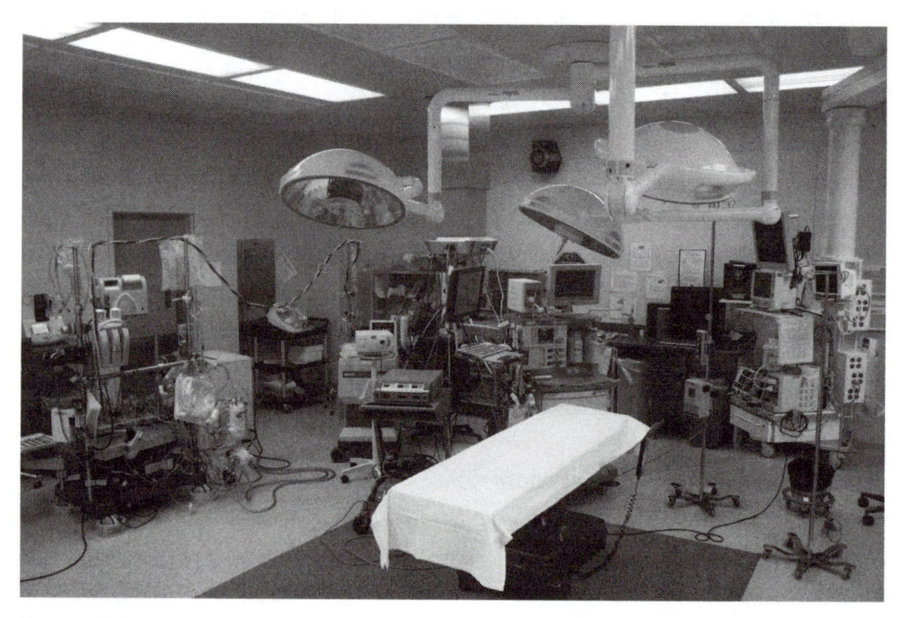

Figure 11.1. Typical operating room.

PART I – THE ENVIRONMENT

Prior to surgery

Before surgery, patients are often placed in a **pre-op area**, also called the surgery prep. Here, patients are often asked to remove their clothes and put on a hospital gown, socks, hair net, and other specialized items. Questions are asked about the person's medical history, allergies, and other information. Some preliminary testing (blood work and urinalysis, for example) may be done. A meeting with the **anesthesiologist** (the doctor who controls the anesthesia delivered to the patient, which make the procedure possible) often occurs at this time.

The operating room

A patient is brought to the operating room by wheelchair or portable bed before being moved onto the operating table. The **operating table** is at the center of most operating rooms and may be surrounded by various other small tables and carts. These hold the equipment and tools needed for the operation. Mounted on the ceiling are bright, specialized lights called **surgical lights**. An example of one type of surgical light is shown in Figure 11.2. These are sometimes controlled by foot pedals so they can be positioned during procedures (also called **surgical cases**). The handle that sticks out of the middle of the bank of lights is usually covered with a plastic hand piece that is sterilized so the staff can move the light during the case.

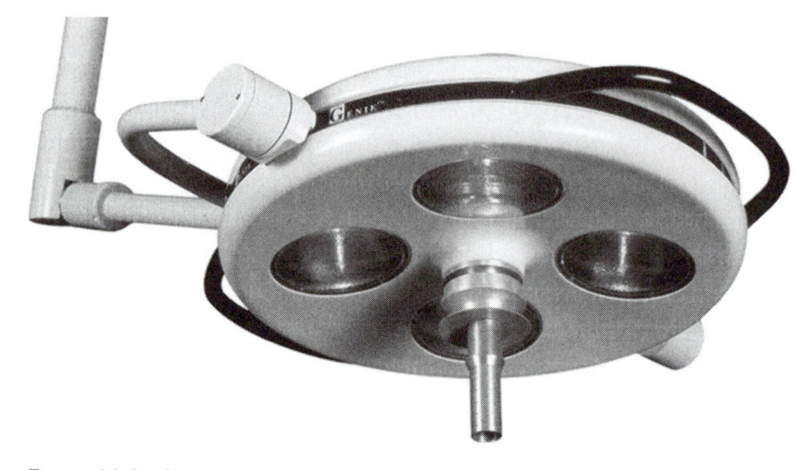

Figure 11.2. Operating room lights.

Anesthesiology

The **anesthesia and monitoring equipment** are kept at the head of the OR table. This is where the anesthesiologist or Certified Registered Nurse Anesthetist (CRNA) delivers medications for anesthesia and monitors the patient during the surgical procedure. Most of the equipment they use is on a cart (sometimes called a workstation). Draeger is one of the main producers of anesthesia equipment.

Once the patient is correctly positioned on the table (there are standard positions used for specific surgeries), medications are used to perform four major functions (not all will be used all the time):

1. **Analgesia** - pain relief
2. **Paralysis (areflexia)** - immobilization (blockage of reflexes like breathing)
3. **Amnesia** - memory loss of events that take place in the OR
4. **Sedation** - deep sleep

Types of **anesthetic agents** (medicines) include gases like nitrous oxide, sevoflurane, desflurane, isoflurane, and halothane. Other drugs (liquid) are also typically given to patients in the intravenous (IV) lines to accomplish some of the four functions previously described. Different anesthetic agents have different actions on the body and are useful in different procedures. Ether gas was a common anesthetic in the past. It is now banned in the United States because of its flammability. Older operating rooms may still contain vestiges of equipment used to prevent explosions.

There are three types of anesthesia:

1. General anesthesia – patients are dependent on a ventilator in this situation, this is "going to sleep"
2. Local anesthesia (often uses novocaine)
3. Regional anesthesia – spinal or epidural

Staff in the OR

Nurses: Two types of nurses work in the operating room: circulating nurses and scrub nurses. **Circulating nurses** prepare the patient for surgery by setting up the IV, attaching the monitoring devices, and helping the anesthesiologist. It is the responsibility of the circulating nurse to prepare the operating room for surgery, set up equipment, and help the scrub nurse place the instruments on the table. During surgery, the circulating nurse sometimes passes items to the surgical team and will also be the one to leave the room if something else is needed or if lab tests need to be performed. Because this nurse is *not part of the sterile field* (the area where the procedure is performed that needs to remain as germ free as possible), she or he takes precautions not to contaminate it. The **scrub nurse** prepares the sterile field, surgical supplies, and equipment. During surgery, the scrub nurse assists the surgeon by passing instruments, suctioning blood, and maintaining the sterile field. Scrub nurses work closely with the operating room team, making sure that everything goes smoothly. After surgery, the scrub nurse or technician washes the "**specials**," which are instruments that personally belong to the surgeon. Some scrub nurses work with only specific surgeons and may be employed by them.

CRNAs: Certified Registered Nurse Anesthetists are specialized nurses who deliver anesthesia. Many BMETs work closely with the CRNAs in a hospital in relation to the equipment in the operating rooms.

Surgical technicians: Also known as surgical or operating room technicians, surgical technicians assist in procedures under the supervision of surgeons, registered nurses, or other surgical personnel. They usually have about one year of specialized training from which they earn a certificate.

The sterile environment

The operating room staff wear special outfits called **scrubs** (a shirt and pants) supplied by the hospital. These clothes are not sterile, but they have been washed under strict guidelines and are presumed to be cleaner than an individual's "street" clothes. (Underclothes and perhaps a T- shirt may be left on, but, for most situations, all other clothes are removed.) Everyone in the room (including BMETs) wears caps, masks, and booties (shoe covers). Facial hair must be covered using a beard cover or sideburn cover.

Workers who are in the sterile field wear additional **gowns** (like a long dress) and rubber gloves. These garments help protect both the person having surgery and the operating room staff against infection and disease. Remember, some diseases (like AIDS) are assumed to be present since a patient cannot be involuntarily tested. *Always assume* that items in the OR have been contaminated with bodily fluids.

BMETs are not usually in the **sterile field**. Unless there is a particular reason, BMETs must avoid entering the sterile field, touching it, or compromising it. (*Note:* The sterile field includes sterile staff. BMETs should never break the sterile field by touching a sterile item or staff member.) BMETs always remain in the nonsterile field (everywhere else in the room). Regulations regarding the sterile and nonsterile fields vary from hospital to hospital and are usually well defined and communicated. Ask questions when unsure of protocol. Following the guidelines of the institution is important for patient and personal safety.

Types of surgery

The four different types of surgery are:

- Diagnostic – This surgery can be for a biopsy or to "see" what an organ or system looks like.
- Preventative – These procedures may prevent a problem before it happens.
- Curative – Patients undergo this type of surgery when a situation can be corrected with a procedure (removing a tumor or repairing a broken bone, for example).
- Palliative – This surgery will enhance the quality of life and can be purely cosmetic or restorative.

Surgical specialties: There are many specialties and subspecialties. For example, a surgeon could operate on pediatric patients with cardiac issues.

- Oncology – cancer treatment
- Orthopedics – bone problems
- Pediatric – young patients, usually under 18
- Neurology – issues of the brain and spinal cord
- Urology – reproductive functions of men, urinary tract in both men and women (kidney and bladder)
- Obstetrics – pregnancy and childbirth
- Gynecology – reproductive system of women
- Plastic/cosmetic – correction to the form of the human body; cosmetic surgery relates to enhancement, and reconstructive surgery is used to correct function
- Cardiac/thoracic – the heart
- Ophthalmology – the eyes
- Otolaryngology – related to the ear, nose, and throat (sometimes called ENTs)

Equipment sterilization

Surgical equipment that is reused must be cleaned and **sterilized** between uses. Items that need to be sterilized include surgical instruments like clamps as well as tools like electric drills. This process can be time consuming and detrimental to the electronics contained inside equipment. Some devices have been carefully designed to be one-time use (disposable) and therefore are not sterilized again.

The most common methods used to sterilize equipment and instruments are

- Autoclave – uses saturated steam at high pressure to kill germs (some are dry heat).
- EtO (ethylene oxide), Sterris is the main manufacturer – uses a gas (harmful to humans) to destroy germs. It does not use high heat and pressure like the autoclave and therefore can be better for some equipment. An example of a sterilizer is shown in Figure 11.3.

Many hospitals designate an area of the hospital in which to process equipment for cleaning and sterilization. Often this area is called Central Sterile, Central Stores, or Central Supply and is abbreviated CS. Staff who work in this area use the sterilization methods described here as well as disinfecting solutions to clean equipment.

Operating room stress and customer service

The operating room is probably the area of the hospital that is the most stressful for the staff. Patients undergo procedures that, even when simple, can have life-threatening complications. Certainly, it is possible that patients who undergo surgery that exposes the heart could have difficulties, but even simple procedures carry risk. As a result, staff depend on the proper and predictable functioning of all equipment all the time. When devices fail to perform as expected, the stress can become visible. It is understandable. BMETs need to provide customer service to the staff with an understanding of the concerns and focus on the patient. Prompt and calm response to a malfunction is critical. Remember that some issues (procedural problems, for example) should be dealt with after the surgical case is completed.

Figure 11.3. Sterilizer.

The fast-paced, high-stress environment of the operating room can cause equipment to fail frequently. Fluids, both from the patient as well as from the procedures, are a constant problem for electronic equipment. Equipment can be accidentally dropped and roughly handled simply because of the focus on the condition of the patient. Also, the cleaning of equipment after procedures, while essential, adds opportunities for damage.

PACU

After surgery, patients are taken to an area called **post-op**, the recovery room, or **post-anesthesia care unit (PACU).** Here, careful monitoring takes place as the patients recover from anesthesia. Nurses and other staff check vital signs, assist with pain relief as the patient wakes up, address surgical site issues (incision care), and generally provide close supervision. Patients eventually are sent to a regular patient room, an ICU bed, or home.

PART II – EQUIPMENT IN THE OR

There are many unique types of equipment used in an OR. Recognize that a BMET's list does not include surgical instruments (clamps and retractors, for example). Most of these are the responsibility of the surgery department. In orthopedics, however, there are some electric devices, such as drills, that might fall into the BMET domain. In addition, many devices found in an operating room are commonly found elsewhere in the clinical setting. Common equipment includes monitors, IV pumps, and defibrillators. Some other types of equipment that may be more specific to the operating room include:

- **Depth-of-anesthesia monitors** – Sometimes called BIS monitors (BIS is a brand name), these devices are able to quantify (on a scale of 0–100) the consciousness of the patient using the EEG. They are discussed in Chapter 9.
- **Anesthesia machines** – These devices deliver anesthetic gases (discussed in Chapter 8). Even though anesthesia machines are fundamentally ventilators, they have some special features. One important function prevents the delivery of more than one anesthetic gas at a time (often there are three or four available). An example of an anesthesia machine is shown in Figure 11.4. You can tell that this machine can deliver three different gases because there are three sets of pressure gauges and three cylinders on the right-hand side of the device.
- **Operating microscopes** – These devices are used to better visualize the surgical field. They are usually binocular and are typically mechanical in nature.
- **Robotics** – These complex devices hold and move laparoscopic instruments. An example of robotics is called the *da Vinci* System. Surgeons use computer-controlled equipment to have finer control of instruments. However, robotics are not found in many ORs because they are useful for only a few specific types of surgical cases, require special training, and are very costly.
- **Smoke evacuators** – These devices are used to remove the material (sometimes called smoke or plume) created by electrosurgical units and LASERs. These common devices may also be called plume evacuators.

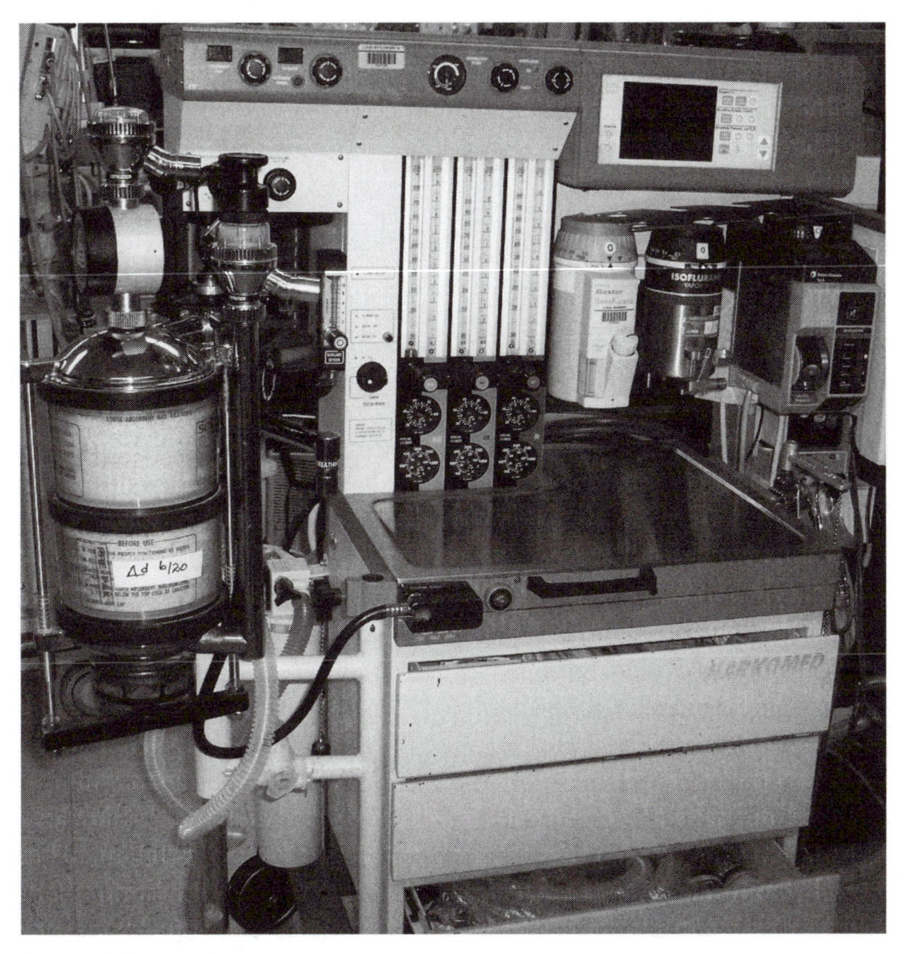

Figure 11.4. Anesthesia machine.

- **Patient warmers** – These devices usually contain warm air that circulates in plastic "blankets." These may also be found in emergency rooms and other parts of the hospital. (The Bair Hugger is a common brand.) Similar devices may also cool patients.
- **Blood and fluid warmers** – These devices, which quickly heat blood to body temperature, often use heated metal

plates or warm water baths. They may also be found in the emergency room.

- **Operating room tables** - Patients lie on special tables that can be moved up and down and tilt. These tables are often controlled by foot pedals.
- **Operating room lights** - These devices need to be very bright and present white light that will not influence the appearance of the patient (shown in Figure 11.2).
- **Saws and drills** - These tools are used to work with bones in orthopedic surgery. They come in many shapes and sizes and are carefully designed for the hospital environment.

Equipment used in laparoscopic surgery

Since the early 1990s, there has been an increase and improvement in the use of minimally invasive surgery. Laparoscopic procedures involve the use of a video camera and a light source, which is put into a patient through a small opening. The surgical instruments are passed through other holes. The Web site http://www.laparoscopy.com contains a great deal of information about these procedures. This site contains many details that are not important to BMETs. The equipment should be the focus of this information.

Scopes: Scopes, which are essentially video cameras, can be either rigid or flexible. Rigid scopes are used for joint surgery and are often called **arthroscopes.**

- Scopes almost always use fiber optics for the light conductor with a xenon lamp as the source.

- The image is almost always seen through a video camera, but it can be viewed through an eyepiece at the end of the shaft.
- Most scopes also have a channel through which fluids, instruments, and gas can be sent into the patient. Many procedures, though, use other incision points to pass instruments into the patient.

Flexible scopes may be called **endoscopes, gastroscopes, proctoscopes, duodenoscopes, bronchoscopes, and cystoscopes**. The part of the body on which these are used is reflected in the name. Because flexible scopes generally enter the body through a natural opening (mouth, nose, urethra), all of the instruments and fluids must be passed through the scope. These devices may also be used outside of surgical cases to examine areas of the body such as the nose or throat.

Trochars: These pointed devices are used to make the hole in the skin to penetrate into the body cavity. They may be spring loaded.

Insufflators: These devices, which are used during laparoscopic procedures, create a gas-filled space within the abdomen in order to move the internal organs out of the way.

Electrosurgical units (ESUs)

These devices are very common in surgery because they are used to assist the surgeon with cutting – that is, making an incision into the tissue (*cauterization*) and limiting bleeding (*coagulation*). ESUs often have two settings, labeled CUT and COAG. Originally, these devices were very basic AC-powered sources that produced waveforms in the RF range 300–3,000 kHz. Today, these ESUs are very complex, microprocessor-driven devices. An

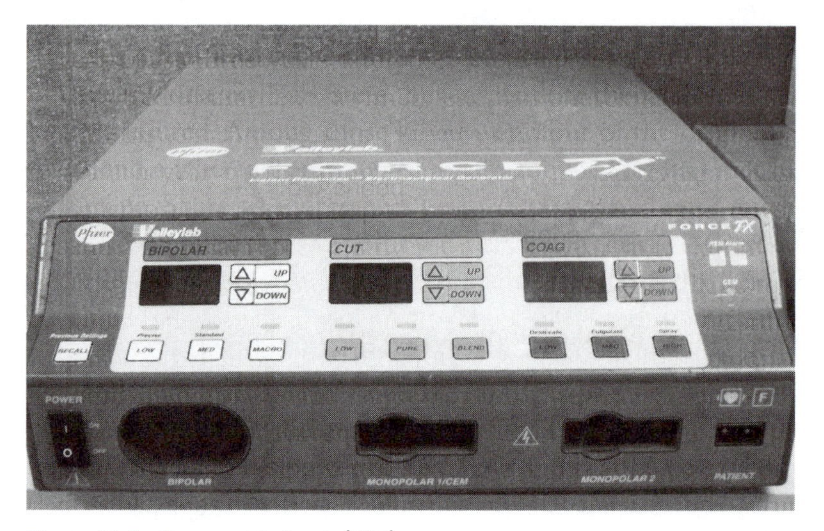

Figure 11.5. Electrosurgical unit (ESU).

example is shown in Figure 11.5. These devices are sometimes referred to as "Bovies" because this is the name of the original inventor, as well as the first brand of units.

The alternative to the use of an ESU is very time consuming, with a less-than-optimal outcome. Without an ESU, all incisions would have to be made with knives, and all bleeding would have to be stopped with sutures to tie blood vessels closed. ESUs are a wonderful technology, but there can be problems - occasionally there are unintended burns to both patients and hospital staff.

Electrosurgical units offer complex and varied waveforms controlled by microprocessors. ESUs operate using one of two energy waveforms. **Monopolar** waveforms use one source of electrical energy. Monopolar devices are often called pencils. A pencil is shown in Figure 11.6. Notice that the pencil has two buttons, one for CUT and one for COAG. This energy travels from the pencil through the patient to the "return electrode" pad. This disposable and sticky pad is filled with electrically conductive gel and returns the energy to the ESU generator. The return electrode is placed on

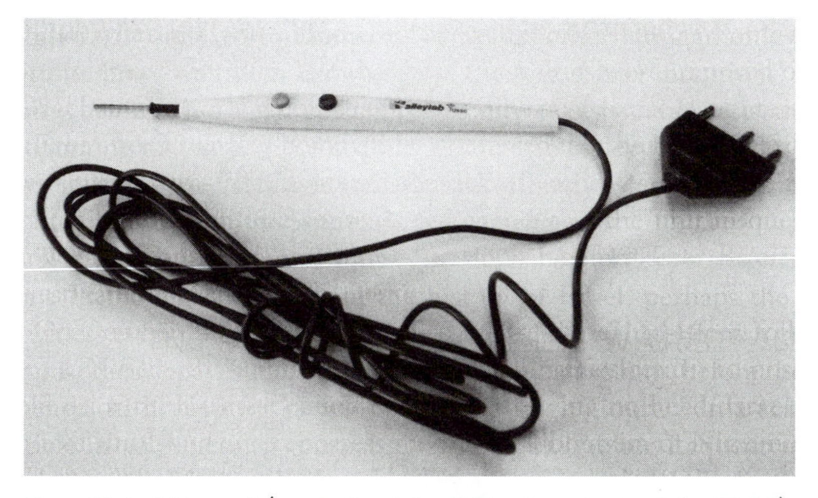

Figure 11.6. ESU pencil (one button is for CUT and one button is for COAG).

the patient in a location that is away from the surgical field. Good contact is important because it is likely that the area used will be covered in sterile drapes and not visible to the staff. Many ESUs maintain a circuit to check the integrity of the return electrode pad - called the return electrode monitor or REM. Should the return electrode become dislodged, the ESU will not work (which can be frustrating to the surgeon during a case).

The second type of ESU energy waveform is **bipolar.** This process passes electrical energy between the two tips of a probe, which looks like tweezers. An example is shown in Figure 11.7. In this form, the electrical energy is very concentrated. The bipolar form is very useful when both ends of the tissue are accessible (such as a blood vessel). However, bipolar cannot be used to create an incision.

Technology used in the cutting and coagulation efforts of surgery have expanded to include the use of LASER light (typically argon) to effect tissue. Also, there are specifically designed waveforms that have very specific effects on tissue. For example, Ligasure fuses tissue or vessels without burning.

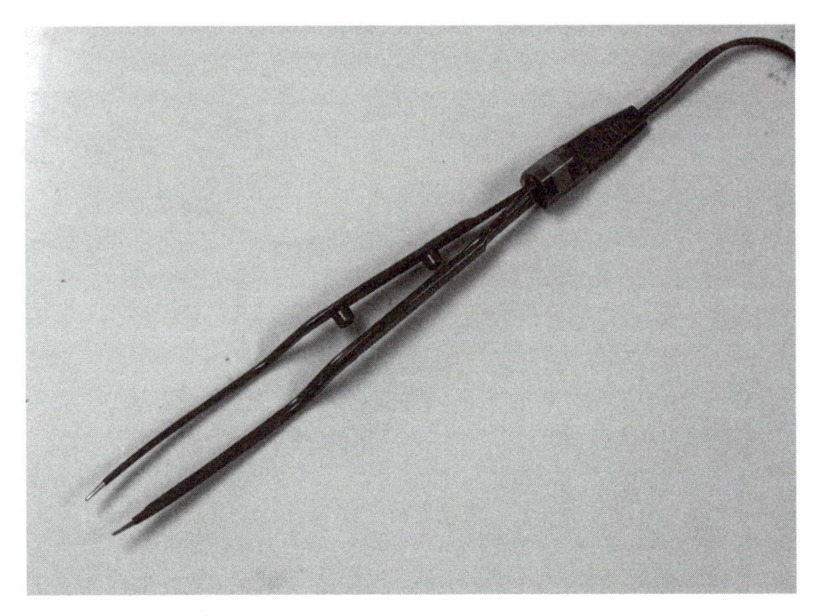

Figure 11.7. Bipolar electrosurgical tool.

As the electrical energy is passed through the tissue, gaseous material (plume) is typically generated. It may be removed from the surgical area by a smoke/plume evacuator.

LASERs

The acronym LASER stands for Light Amplification by the Stimulated Emission of Radiation. This device uses specially treated light to alter tissues in a specific way. LASERs typically coagulate or vaporize (cut) tissue.

In general, LASER light waves are:

- Coherent - All of the light waves are in a single phase.
- Monochromatic - All of the light waves are of a single color (frequency).

- Collimated – The light waves are capable of staying together in a tight beam over long distances.

Some LASERs are capable of producing more than one color and produce different tissue effects. LASER light can be made from four different sources (mediums): solid, liquid, gas, or electronic.

Some common types of medical LASERs are:

- CO_2 (carbon dioxide) – very common surgical LASER for cutting and coagulation
- Nd:YAG (neodymium yttrium aluminum garnet)
- Tunable dye
- Argon
- Ruby

Some functions of LASERs include corrective eye surgery, removal of tumors, removal of tattoos, and skin resurfacing.

STUDY QUESTIONS

1. Describe the responsibilities of anesthesiologists. Include the technologies used by anesthesiologists to support patients.
2. Describe what a BMET would wear to enter an operating room.
3. What are the types of surgical specialties? Create a list of each type. Flash cards may assist in this effort.
4. Identify and describe some of the customer service qualities that are important for BMETs in the OR.
5. Identify and describe some of the environmental pressures on equipment that are specific to OR equipment.
6. Make a list of common equipment you would find for a typical case (not laparoscopic or robotic) in an operating room.

7. Identify the type of electrosurgical technique that requires a return electrode. Sketch the electrical path (from the ESU - back to the ESU) through the patient.

FOR FURTHER EXPLORATION

1. Use the Internet to research OR patient tables. Document their characteristics. Describe the features that make them different from the beds used in patient care rooms. Identify and describe some of the types of OR patient tables that are designed for specific types of surgery.
2. Use the Internet to research the use of irrigation during surgical procedures. Discuss how this impacts the surgical field. There has been a great deal of debate as to whether ORs should be defined as wet locations by NFPA (currently, they are not). In addition, Cesarean sections, which deliver infants, release a great deal of amniotic fluid. Discuss how this too contributes to the wetness of the surgical area. Discuss how technology and staff within this area must be aware of the impact of the fluids.
3. Research and document some standard patient positions with names such as Sims, Fowlers, prone, supine, lithotomy, and Trendelenburg.
4. Research and describe the steps required in the EtO (most common) sterilization process for equipment reuse.
5. Research and summarize the characteristics of orthopedic equipment such as drills used in surgery. How do they differ from the kind of tools found in a hardware store?
6. Robotic surgery is very popular in the media. Research the components of robotic surgery and summarize them. Describe some advantages and disadvantages of robotic use in surgery.

7. Explosions were a constant threat with the use of ether. Define and describe ether as an anesthetic gas. Explore and summarize historical information to find out what precautions were taken to avoid dangers in operating rooms that used ether. Be sure to discuss such precautions as flooring restrictions and electrical plug configurations. If possible, include the testing required of BMETs to ensure safety in operating rooms that used ether.
8. List and describe the equipment needed for a laparoscopic appendectomy. Describe the advantages to the patient of a laparoscopic appendectomy instead of a traditional procedure with an abdominal incision.
9. When patient illustrations are viewed at http://www.laparoscopy.com, patients often appear very large and bloated. Why is that? Specifically identify the equipment that causes this condition. Explain why this inflation is necessary.
10. Valley Lab (a major manufacturer) produces good reference materials. Use the following site as a useful resource: http://www.valleylab.com/education/poes/index.html (Review the Principles of Electrosurgery link on the left). Define and describe the following concepts:
    - ▷ bipolar and monopolar mode
    - ▷ patient return electrode use and placement
    - ▷ REM (return electrode monitoring)
    - ▷ basic setup and function of the use of an ESU
11. Identify and describe the potential life-threatening medical complications that could compromise patients undergoing even the most simple and short surgical procedure. (*Hint:* Search the Internet for *surgical complications.*) Identify and describe potential equipment-related hazards for patients (include hazards from LASERs and ESU devices).

12. Cardiac surgery often involves specialized equipment not described in this chapter. Examples include bypass machines and pacemaker programming devices. Research, identify, and describe these devices. Identify other devices used in cardiac surgery.
13. Visit this Web site, http://www.streamor.com/orintro/orintroindex.html, to view streaming video of an OR. The Web site contains a great deal of information about the staff in an OR as well as what *scrubbing* means. Prepare a summary of:
    - ▷ people in the OR
    - ▷ equipment in the OR
    - ▷ scrubbing for the OR
    - ▷ sterile techniques

12

Imaging

LEARNING OBJECTIVES

1. describe ultrasound imaging
2. describe basic x-ray imaging
3. describe CT scanners and identify how they differ from basic x-ray devices
4. describe DICOM and PACS
5. describe and characterize MRI
6. describe and characterize nuclear medicine – both imaging and treatment
7. describe and characterize PET scans

Introduction

The technology involved in imaging the inside of the human body is tremendous and amazingly broad. There are a vast array of unique devices, but this chapter will deal with general categories (based on the technique or source of image) of imaging devices.

Ultrasound

Ultrasound uses sound waves to image anatomy and anatomical function. Ultrasound can also be used to detect blood flow, anatomic movement (heart valves, for example), and, most commonly, anatomical structures. It is frequently chosen as an imaging modality because it does not use radiation and carries less risk to the patient. Ultrasound uses high-frequency sound waves (1–20 MHz) that are projected into the body. The waves hit an object, bounce back, and are detected with a piezoelectric crystal sensor. The distance from the transmitter probe and the anatomical structure can be calculated, and this data is used to produce an image. Computers can manipulate the transducer data to present three-dimensional or moving images.

Transducer probes, which contain both the transmitter and the sensor, come in many shapes and sizes. Some are designed to enter body cavities such as the esophagus and the vagina for better imaging quality. Most commonly, probes are placed on top of the skin and transmit into soft tissues. Ultrasound waves are blocked by air and bone. This can be a problem since air is contained in the lungs and there are bones surrounding some organs like the brain and heart. A gel is often used to help block any air between the probe and the skin. This ensures effective transmission of the

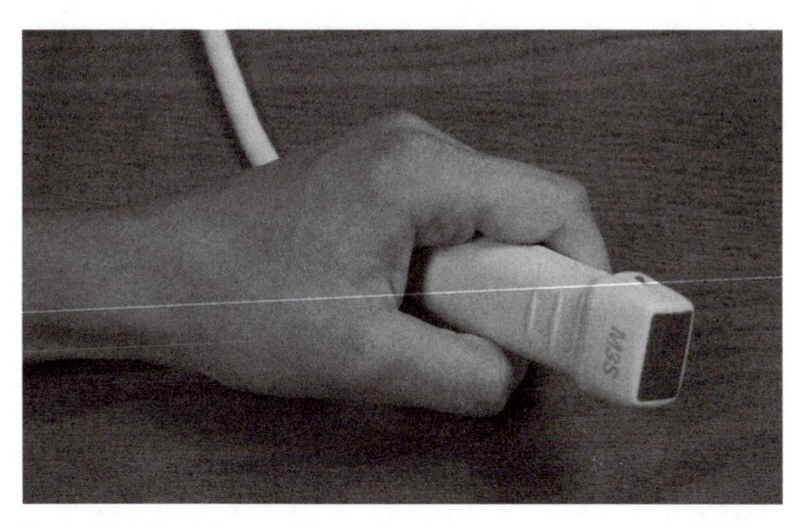

Figure 12.1. Trans-thoracic ultrasound probe.

waves and unobstructed images. An example of a probe used for a trans-thoracic (cardiac) ultrasound is shown in Figure 12.1.

Because of the inherent safety of this type of imaging, women often undergo ultrasound imaging to evaluate the progress of pregnancy. Many infant abnormalities can be detected early and, in some cases, perinatal treatment can take place. Another common application of ultrasound is echocardiography. The anatomy and function of the heart can be examined. The data from the transducer is paired with ECG information to provide the cardiologist a great deal of heart performance information. A portable ultrasound machine is shown in Figure 12.2.

X-rays

Radiation comes in many wavelengths, a common one is the x-ray. The x-ray wavelength evolved to become the name of a

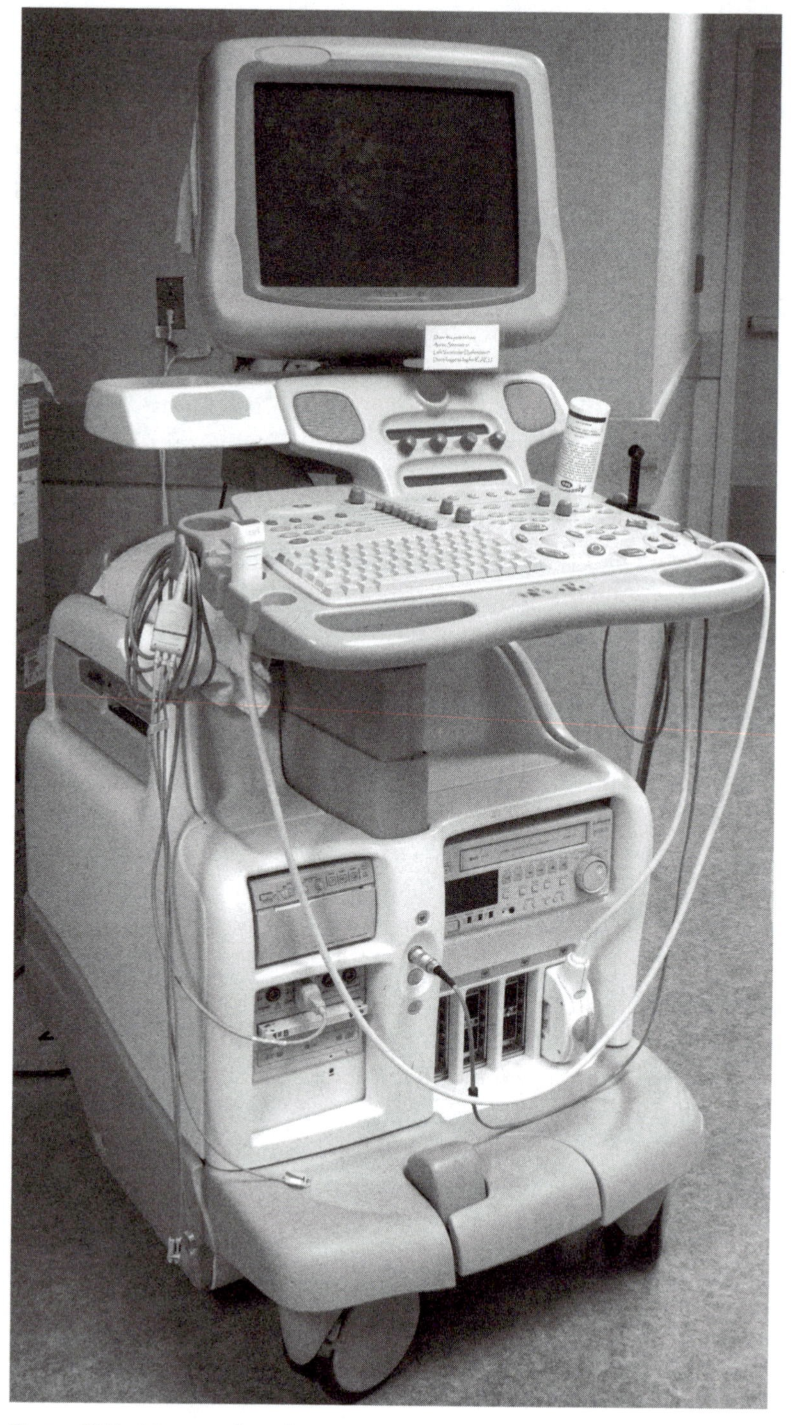

Figure 12.2. Ultrasound machine.

device, a machine that uses this wavelength of radiation. A basic x-ray machine, used to check on a broken ankle or arm, has a long history and serves a useful purpose. X-rays are blocked by different materials at different levels, including the calcium in bones and teeth. An image can be created in a fairly basic way. The machine uses a source of x-rays (in a carefully controlled housing) and a type of film that is altered by x-rays. The patient (who is a source of x-ray blocking materials such as bone) is placed in between the source of x-rays and the film. Consequently, the image formed shows exposed film surrounded by a shadow of the bone of the patient. This type of imaging is commonly used in dentist offices and hospitals and can simply diagnose many diseases and injuries. X-ray imaging can be adapted for more specialized medical uses. For example, **mammography** is simply a specialized x-ray for the breast. X-rays are also used in CT devices and fluoroscopy machines.

Computed tomography or computerized axial tomography

The computed tomography (CT) or computerized axial tomography (CAT) machine is shaped like a large donut. As the patient lies in the center of the CT unit, an x-ray tube circles the patient, sending x-rays into the patient at many locations. Multiple detectors pick up the radiation. A computer combines the many individual x-ray images into cross-sectional pictures, or "slices" of the body. The absorption characteristics of the radiated tissues (especially with the use of contrast agents injected into the body) and the intensity of the x-ray beam produce images showing internal structures such as blood clots, skull fractures, tumors, and infections. The ability to image soft tissues (compared with the basic x-ray system) was a tremendous advancement in medical

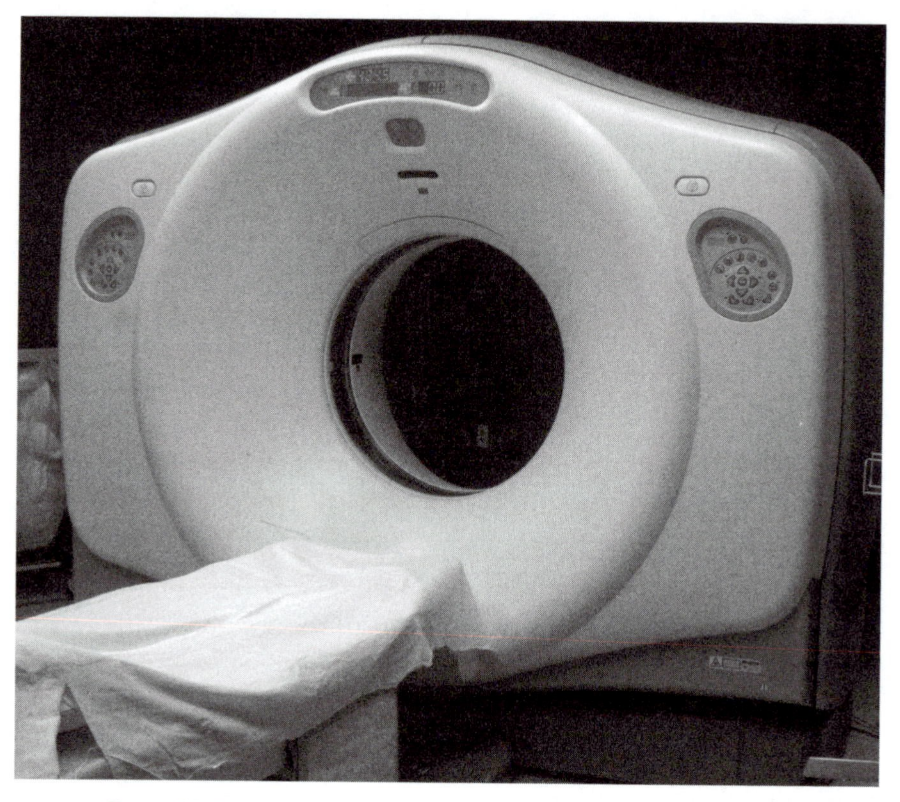

Figure 12.3. CT scanner.

imaging. Basic CT images are typically cross-sectional views of the body (similar to a slice of bread as if the loaf were the human body). As CT technology has integrated more techniques, the images generated from the many sources and detectors can be manipulated to create rotatable, three-dimensional images or three-dimensional plastic models. A common CT scanner is shown in Figure 12.3.

Fluoroscopy

This real-time x-ray imaging system produces images on a video camera (either analog or digital). Essentially, the x-ray images are able to continuously change as the patient (or something inside the patient) moves. Fluoroscopy, for example, can track the movement of a chemical through the digestive tract as a patient swallows. Fluoroscopy got its name from the use of fluorescent screens that previously displayed the moving image before the use of cameras. Angiography, a specific use of fluoroscopy, allows physicians the ability to look at blood flow through the blood vessels using fluoroscopy. Cardiac catheterization and angioplasty use the real-time images to view coronary blood flow and treat blockages. **C-arm** devices are fluoroscopic x-ray machines, which can be portable or fixed. Many times these are digital devices (producing a digital image rather than using film) and are useful in many applications including during surgical cases. A c-arm is shown in Figure 12.4.

Common applications of x-ray imaging

Cardiac catheterization: In this procedure, a doctor guides a catheter through an artery or vein in the arm or leg, into the heart, and then into the coronary arteries in the heart. The catheter is visible to x-rays to guide the insertion. Using specialized chemicals, the procedure can measure blood pressure and how much oxygen is in the blood and can provide other information about the pumping ability of the heart muscle or for treatment. The process is guided using moving x-ray images. When a catheter is used to inject dye into the coronary arteries, this is termed coronary angiography or coronary arteriography. If a catheter has a balloon on the tip, the procedure is known as percutaneous transluminal coronary angioplasty (PTCA).

Angioplasty: PTCA is a procedure used to dilate (widen) narrowed arteries. Using the guidance of x-rays and a fluoroscopic

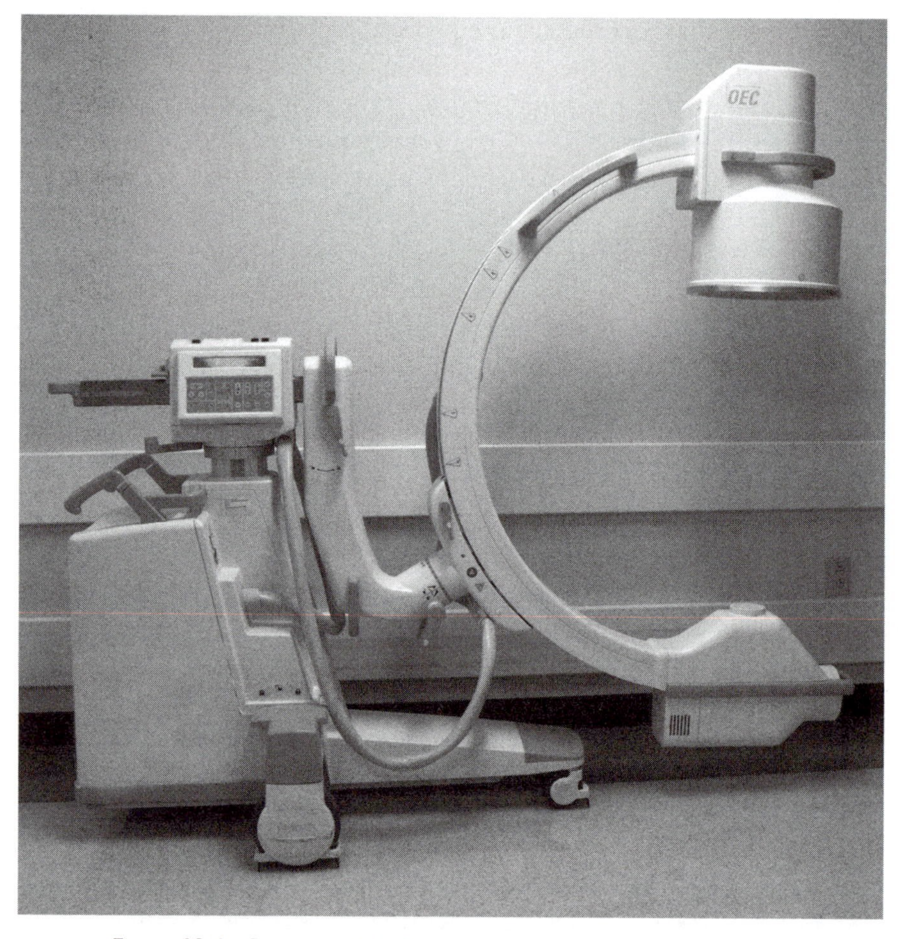

Figure 12.4. C-arm portable fluoroscopy unit.

camera, a doctor inserts a catheter with a deflated balloon at its tip into the narrowed part of the artery. Then the balloon is inflated, compressing the plaque and enlarging the inner diameter of the blood vessel so blood can flow more easily. Then the balloon is deflated and the catheter removed. LASERs are often used to assist in the widening of arteries, and stents are frequently placed in the artery during this procedure.

Contrast medium, or, more simply, "contrast" may be added to specific areas of a patient (soft tissue structures such as blood vessels or parts of the urinary or digestive system) to enhance the x-ray images. The solution changes the way some tissues and organs appear to x-rays. Sometimes contrast agents may be called "dye" but contrast medium is not a colored dye. To enter the body, contrast may be swallowed or injected (through a catheter), inhaled, or inserted into a body cavity.

Magnetic resonance imaging

Magnetic resonance imaging (MRI) uses magnets and radio waves (not radiation). First, the subject is placed inside a tremendously high-powered electromagnet (0.2–3 tesla (T), although this is constantly changing as research and design expand MRI applications). This causes the hydrogen atoms inside the human to align with the north and south poles of the magnet. Then pulses of radio waves (at a wavelength designed to excite just hydrogen) are delivered to the patient. This causes the hydrogen atoms to spin. As they come to realign with the magnets, they give off radio waves. The radio waves can be picked up by a detector and the information about the hydrogen atoms can be processed by computers. Molecules such as glucose are made up of hydrogen atoms. The processing of this data provides information about how organs use the glucose, for example. Uniquely, MRI allows images of not just the liver but how the liver is functioning. Fundamentally, MRI images the chemistry of the body. This is very different from other imaging devices. Normal and abnormal tissues of the same organ appear different to this imaging device. In addition, MRI imaging is able to see some parts of the body that

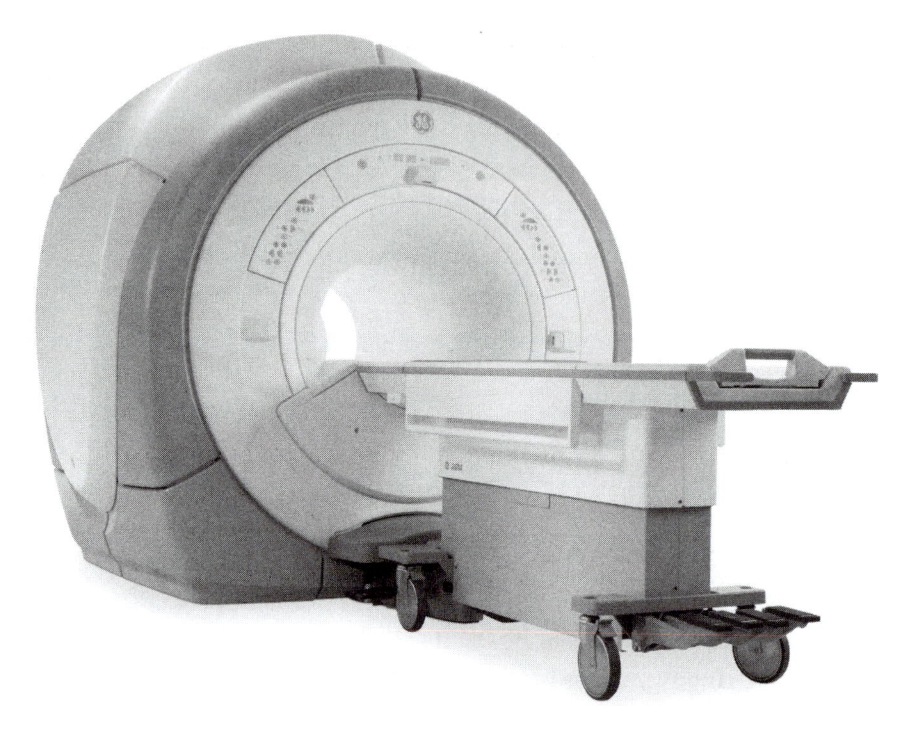

Figure 12.5. GE MRI machine. (Photo courtesy of GE Healthcare.)

are not clearly visible to CT scans. An example of an MRI device is shown in Figure 12.5.

Safety and MRI

Safety must be a top priority when in the presence of an MRI machine because of the extreme power of the magnet used. There are interesting precautions that must be taken both in the room that houses the MRI as well as for the patients. No metallic objects can go near the magnet. Screwdrivers can be ripped from the pocket of a BMET if the magnet is activated. Patients cannot have pacemakers or many kinds of implants (although some embedded in bone may be allowed), and even some metallic inks used in tattoos can be a problem. In addition, patients often need to be monitored and connected to ventilators and IV pumps while an MRI is performed. Devices must be carefully constructed to shield the metallic components from the powerful magnet and radio waves.

Nuclear medicine

Nuclear medicine uses very small amounts of radioactive chemicals to view (and treat) the human body. The use of these materials can be helpful to image parts of the body (check for tumors, for example) or to target known areas of disease. Gamma radiation (a specific wavelength radiation) is used, for example, to scan the bones. The patient is injected with a very specific chemical, which is attracted to the bone and emits radiation. A specialized gamma camera detects the radiation emitted from the patient. The radioactive chemicals are excreted from the body quickly.

Positron emission tomography

PET is an imaging modality that is able to image organ and tissue function with great clarity, and PET scans are able to image tissue metabolism. A compound, usually sugar, is specially altered to release positrons (an isotope). This special sugar (FDG) is injected into the patient. The sugar is transported around the patient's body. The isotope reacts with sugar inside the patient to produce two gamma rays 180 degrees apart. Inside the scanner, a ring of detectors look for the gamma rays and can determine the location of this special compound in the body. Computers can correlate all the emission data to map tissues in the body. PET scans have relatively common use in cancer diagnosis because cancer cells use sugar in high quantities. Currently, PET scans are combined with CT scanners to produce an anatomical image in

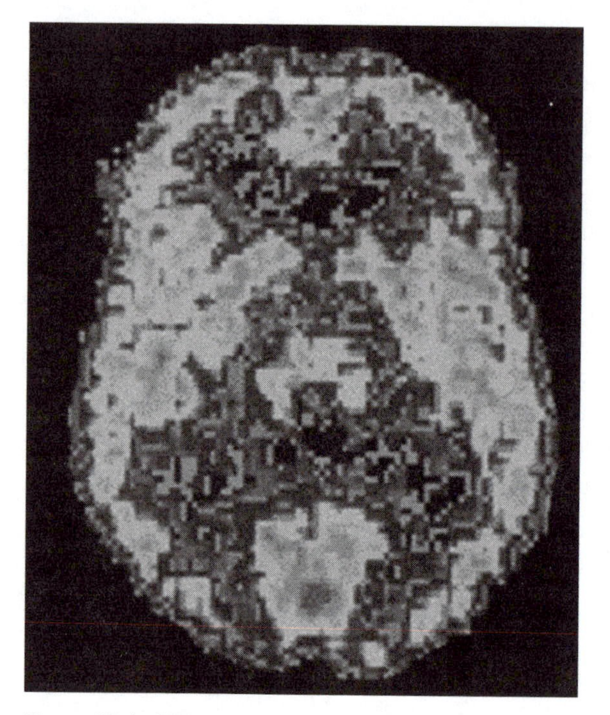

Figure 12.6. PET scan image.

combination with the sugar metabolism information. An image is shown in Figure 12.6.

The isotope is produced using a cyclotron and has a half-life (amount of time during which half of its radioactivity is diminished) that limits the distance that the isotope can be transported (essentially restricted to the distance between the scanner and the cyclotron). Commonly used FDG has a half-life of about 2 hours, so a careful transportation network is in place to deliver the compound to the scanner site at the right time. Other compounds can be used for specific applications like cardiac studies and bone examinations, but these compounds, such as carbon and nitrogen, have such short half-lives that the practicality of the scans is currently a challenge.

Picture archiving and communication systems

The software and technical specifications used to store digital images and, more importantly, access the images on a variety of devices is vital to the effective use of digital images. Traditionally, images were saved on plastic film. A shift in "film" storage is occurring as images are being created in digital form. These images can then be networked, emailed, and transmitted to many locations without the plastic sheet being carried from one place to another. This storage is called **PACS,** which stands for picture archiving and communication systems. There is a vital interrelationship between devices, image access, medical device manufacturers, and network computer technology.

DICOM

Digital imaging and computing in medicine (DICOM) is the **standard** (the rules) that allows connectivity between the devices, which acquire the image, and the computers, networks, printers, and servers, which share them. This is the agreed-upon file format and communication protocols that enables all equipment to work together.

STUDY QUESTIONS

1. Name the two types of imaging modalities most commonly used to observe organ motion.
2. Identify the main differences between a traditional x-ray device and a CT scanner.
3. Define a contrast agent and describe how they are used to enhance images.

4. A small permanent magnet (such as a decorative magnet) has the strength of about 0.01 T. Compare this to the magnet used in MRI devices.
5. Describe some of the practical safety concerns required for work in MRI rooms.
6. Identify which devices can image organ function.
7. List and describe imaging modalities that use radiation.

FOR FURTHER EXPLORATION

1. Research, list, and describe five measurements taken during a fetal ultrasound procedure that can indicate abnormalities. For example, the distance between the eyes can indicate Down syndrome. Explore and document what conditions can be detected during fetal ultrasound.
2. Research, define, and describe ejection fraction and cardiac output, two numerical values that can be determined from echocardiography. Identify normal values. Describe abnormal values and the disease conditions they may indicate.
3. Real-time videos taken using echocardiography are vital to the cardiologist. Visit http://www.echobasics.de to view real-time videos of various parts of the heart. Select the section labeled "transthoracic examination." Research and define a transthoracic ultrasound. Describe how the test is performed. When exploring the Web site, observe the valves opening and closing.
4. Basic x-ray machines use a Bucky to guide the x-rays. Research, define, and describe this device. Document its function within an x-ray machine.

5. Research, document, and explain at least five types of diseases and conditions that can be detected simply with a basic x-ray machine. Include both injuries and diseases.
6. Explore the Web site http://www.ptca.org/devices2.html, which contains interesting photos of fluoroscopy imaging equipment. Identify and document the overall purpose of the procedure described. Define and describe "balloon" and "stent" as described on this Web site.
7. http://www.sprawls.org is an online textbook about medical imaging. The text is called *Physics and Technology of Medical Imaging*. Click on *Physical Principles of Medical Imaging* to view the Table of Contents. In the first module, *General Medical Imaging Topics*, read the *Medical Image Characteristics and Quality Factors* section. Define image quality. Identify the five image quality characteristics and define and describe each of them.
8. Research, define, and describe the ratings that are used on equipment for use in the MRI room – **MRI safe** vs. **MRI compatible.** How do the ratings differ?
9. Physical size of a patient is a problem for some imaging devices that must encircle the patient. CT scans and, more commonly, MRI machines do have a limitation in the opening size through which a patient must fit. Claustrophobia can also be an issue. Research, document, and describe both physical size limitations and claustrophobia. Describe the impact of these situations on patient care.
10. CT and MRI machines are very expensive. Document and describe how the community surrounding a hospital might impact the number and type of devices of this cost a hospital can purchase. Research local hospitals to determine how many of these devices they own, and describe how these statistics might vary from rural to urban areas.

11. The journal *Medical Imaging* is available on the Internet at http://www.medicalimagingmag.com/. Visit this Web site and summarize an interesting article.
12. Research the amount of radiation exposure (measured in Sieverts) typical for a chest x-ray, a CT scan of the chest, and a dose of radioactive material such as PDG during a PET. Compare the dosages and discuss the impact of radiation exposure to patients for different imaging modalities.

13

Clinical laboratory equipment

LEARNING OBJECTIVES

1. describe and characterize the purpose of the clinical laboratory
2. describe Coulter's discovery
3. define flow cytometry and describe why it is useful
4. describe and define a fluorochrome and identify why it is useful
5. describe potentiometry
6. define an ion selective electrode and describe its applications
7. define spectrophotometry and describe the application of this technique
8. define osmolality and describe the application of this technique
9. define an osmometer and identify its basic principle
10. describe electrophoresis
11. describe the principles of operation of a centrifuge
12. identify and describe other types of laboratory equipment
13. describe point-of-care testing

Introduction

The clinical laboratory is vital to a hospital's ability to treat patients. Bodily fluids from patients are analyzed for their chemical makeup and evidence of disease. Many therapeutic and treatment decisions are made based on the results of this important analysis. Almost all testing is automated and most devices perform many tasks. BMET support is vital to the clinical laboratory, as it is a very equipment-intensive department. The high level of complex equipment does require some hospitals to obtain service contracts for the support of clinical laboratory equipment.

Some of the fluids from patients that are commonly analyzed include blood, urine, and cerebral spinal fluid. A tremendous number of tests are performed. This chapter focuses on the principles that form the foundation for the complex equipment. Most devices use multiple reagents, bar codes, robotics, and computers to perform the sample preparation and multiple tests.

In addition to automated fluid analysis, many laboratory tests evaluate tissue and fluid samples for evidence of abnormalities. Many of these tests are not automated and require the expertise of a pathologist or clinical laboratory technician.

It is worth noting that the terms "medical technology" and "clinical technology" may be seen interchangeably, although medical technology is an older label.

Clinical laboratories have their own specific regulations and organizations that offer guidelines and best practices. These include the American Association of Blood Banks (AABB), College of American Pathologists, and Clinical Laboratory Improvement Amendments (CLIA). CLIA are federal requirements.

Blood analysis

Many tests are performed on blood. A common test is blood cell count. The automation of this test was vital to improvement in analysis speed. W. H. **Coulter** created an automated method of particle counting using electrical impedance. As the blood cells pass between electrodes, the impedance increases in pulses. These pulses can be counted to determine the amount of cells (particles) in a sample. Since red blood cells and platelets have different sizes, the pulses vary in size. This allows for very specific results.

Blood sample analyzers combine the Coulter technique with other technologies to provide additional information and/or faster speed. For example, to determine the amounts of each of the five different white blood cell types, flow cytometry and radio frequency signals are used.

Flow cytometry uses laser beams and photodetectors to measure the light that passes through cells as well as the light that is reflected. Fluorescent dyes are commonly used to assist in the specific cell analysis. **Fluorochromes** are chemicals used to identify antigens or antibodies because they are able to absorb and reemit light. Flow cytometry assists with the identification and sorting of different types of cells in a moving liquid.

Another important quality of blood is the dissolved blood gases (carbon dioxide and oxygen) and the pH of the blood (normal blood pH is 7.35–7.45). **Potentiometry** is a technique that uses electrodes to detect the quantity of these substances in a sample by measuring the voltage between the electrodes. This has become a very useful technique as **ion selective electrodes** (ISEs) have been created. These electrodes are carefully designed to measure the quantity of specific chemicals including glucose, potassium, fluoride, lead, mercury, and calcium. This technique only requires very small samples, on the order of microliters. Scientific experimentation that creates new ISEs is ongoing and expands the substances that can be evaluated using this technique.

The use of light in laboratory tests: A majority of clinical lab measurements use some type of **photometry**, which simply means light measurements. Photometry plays an important role in some of the techniques described earlier. **Spectrophotometry** is a technique that exposes a fluid sample to a light beam and measures the amount of light transmitted through the solution.

Urine tests

Osmolality is the measurement of the concentration (amount) of particles in solution. The "solution" is the urine sample. The "particles" to be measured are elements such as electrolytes (sodium, potassium, calcium), glucose, drugs, and antibiotics. The instrument used to do this is an osmometer. The science of measuring osmolality is called **osmometry**.

Osmosis is the process of moving water across a semipermeable membrane (the cell wall) until there is an equal concentration of fluid on both sides of the membrane. This is an essential process for moving nutrients, drugs, and antibiotics into a cell as well as allowing waste materials to pass out from the interior of the cell.

The **osmometer** is a laboratory device that measures the osmolality of a solution. Virtually all osmometers in clinical laboratories determine osmolality by measuring freezing point depression, which is a chemical property that is proportional to the amount of compound in the sample (this principle states that the freezing point of a liquid is lowered in proportion to the amount of chemical present).

The patient's specimen is poured into an analysis vial and a thermistor probe is used to record temperatures of the sample. Then the sample is cooled and warmed several times during which temperatures are recorded. This data is used in the electronic calculation of the osmolality of the patient's specimen. This data can be compared to known concentrations of compounds to determine the make-up of the particular sample.

Electrophoresis

This technique is used to separate large molecules such as proteins. Applying electrical charge causes molecules to separate based on weight. Placing the sample in a gel can lock molecules in place after they are moved by the electrical field. Stains may also be used to assist in the visibility of molecules in the gel. Samples that contain DNA strands can be "decoded" to assist in genetic identification. DNA pattern matching can be used to determine tissue sample matches (in forensic investigations) and identify familial relationships. The resulting patterns in the gel are often shown as visible representation of DNA patterns. Bands are displayed in Figure 13.1.

Centrifuges

Many automated devices have centrifuges built into them; nevertheless, stand-alone centrifuges are often found within the clinical lab and in many other locations throughout the hospital.

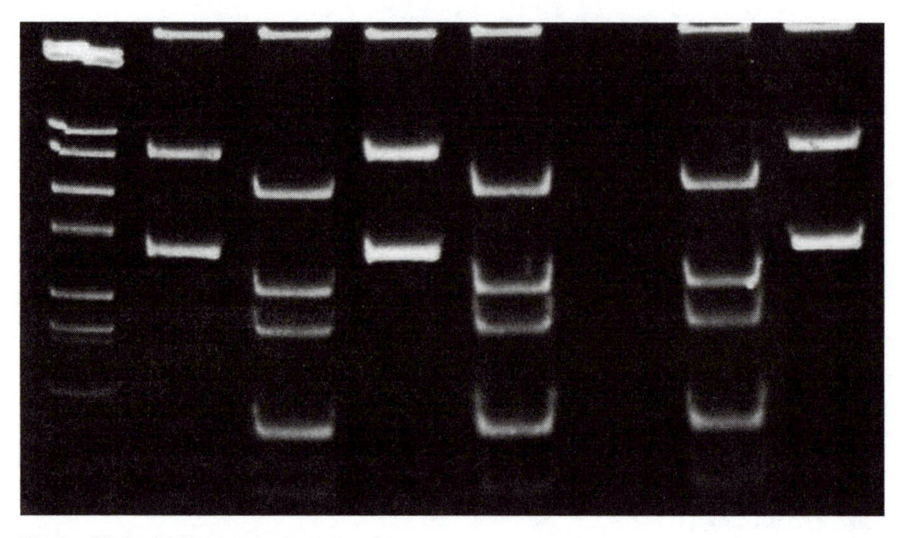

Figure 13.1. DNA electrophoresis gel.

Many medical facilities like doctors' offices have centrifuges as well. A centrifuge is a device that spins samples to separate the particles from the solution according to their size, shape, and density. Centrifugal force separates the particles.

A centrifuge is basically an electric motor used for spinning samples of fluid for further analysis. It often contains test-tube-shaped wells for holding specimens. A typical centrifuge is shown in Figure 13.2.

A centrifuge spins because of the principles that guide motor action:

- Current flows through the **field coils** (also called stator)
- Which generates a **stationary magnetic field**
- Which generates a magnetic field on the **armature** (also called a rotor)
- Current flows from the brushes onto the armature **commutator**

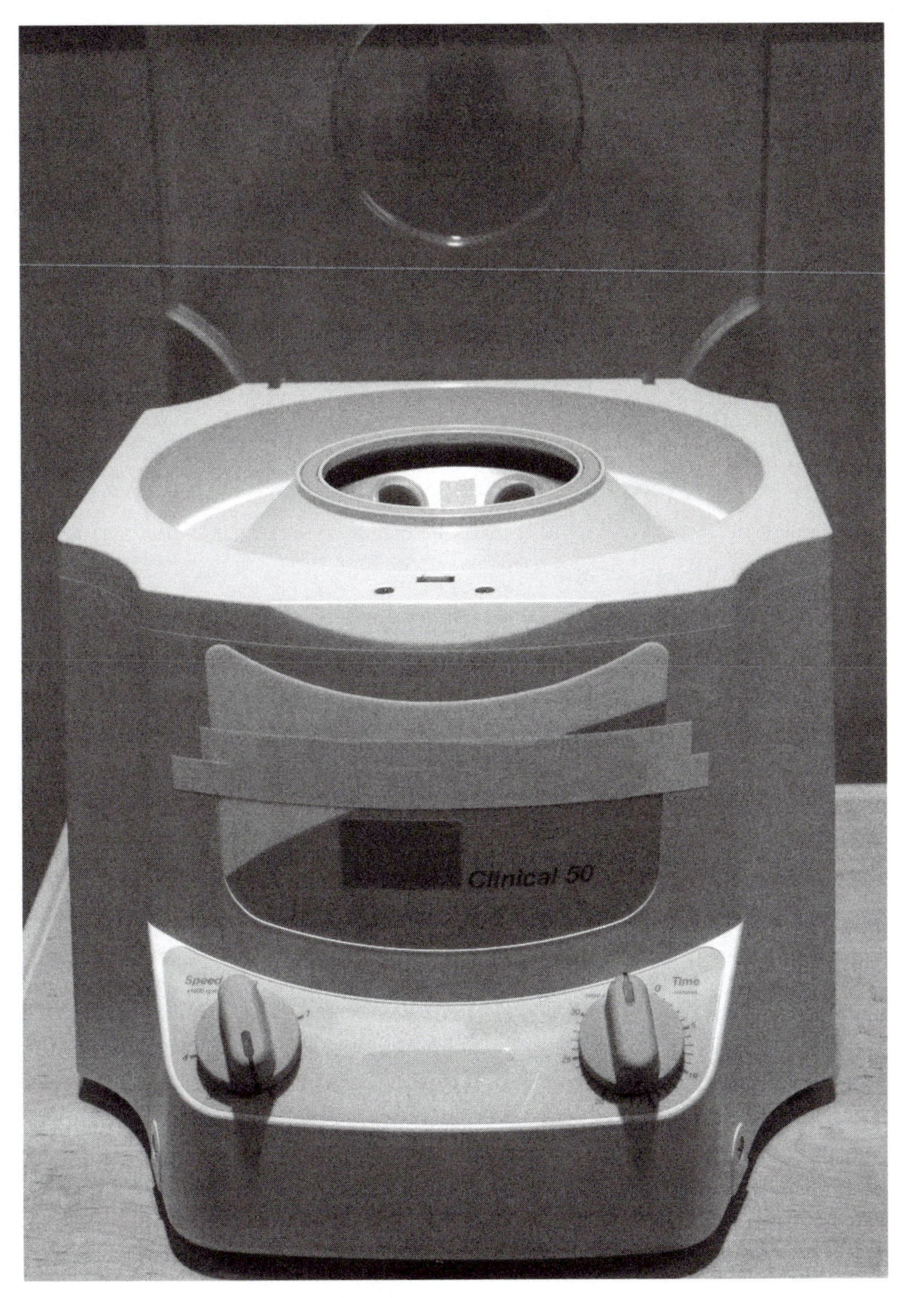

Figure 13.2. Centrifuge.

- Current flows through one of the armature coils, which generates a magnetic field, which is 90 degrees out of phase with the stationary field
- Magnetic force tries to align the armature coil
- As soon as the armature coil starts to move, contact with the brushes is broken and there is no longer a magnetic field in the armature coil
- Meanwhile this process is repeating with another armature since it is now in contact with the brushes and that armature is attracted to the stationary field

As the process of creating an electromagnet and movement (which destroys that electromagnet) is repeated, the armature continues to rotate.

Speed can be controlled by varying the voltage sent to the field coils and armature that adjusts the strength of the magnetic field. Spin is dependent on a balance of samples placed inside the device (one sample opposite to the other).

Very commonly, the **brushes** suffer mechanical wear and must be replaced.

When working on centrifuges, recognize that there is an inherent risk from contamination with bodily fluids that have spilled or splashed when sample holders are not correctly sealed or break.

Other clinical laboratory equipment

Microscopes: These magnification devices are often more complicated than the typical light microscope many students have used during their education. Laboratory microscopes may contain a digital camera or computer interface.

Microtomes: These specialized tissue slicers "section" (slice) samples into very thin pieces for better viewing under a microscope or for further testing. Some of these instruments are mechanical, but others can be very complex.

Tissue stainers: In an effort to streamline laboratory tests, automated tissue stainers not only apply stain to samples in uniform amounts but also agitate the samples, heat them, or perform other actions.

Laboratory-grade refrigerators: Many samples (and blood in the blood bank) must be kept at precise temperatures to ensure accuracy and safety. Laboratory refrigerators are able to do this. They operate under the same principles as a household refrigerator, simply with better temperature monitoring and control. Refrigerators often have built-in alarms to notify personnel of errant temperatures.

Point-of-care testing

To streamline patient care and speed up the time between sample collection and treatment decisions, some common tests can be performed at the patient bedside. Specialized devices or modules are often integrated into patient monitors and can perform a wide variety of tests. Blood gases, electrolytes, and hemoglobin, for example, can all be tested at the patient bedside. The dramatic change in the time needed to obtain test results has greatly enhanced patient care. Figure 13.3 shows a Philips monitor with a rack below that supports modules. Many of the modules are specific for a type of bedside laboratory test.

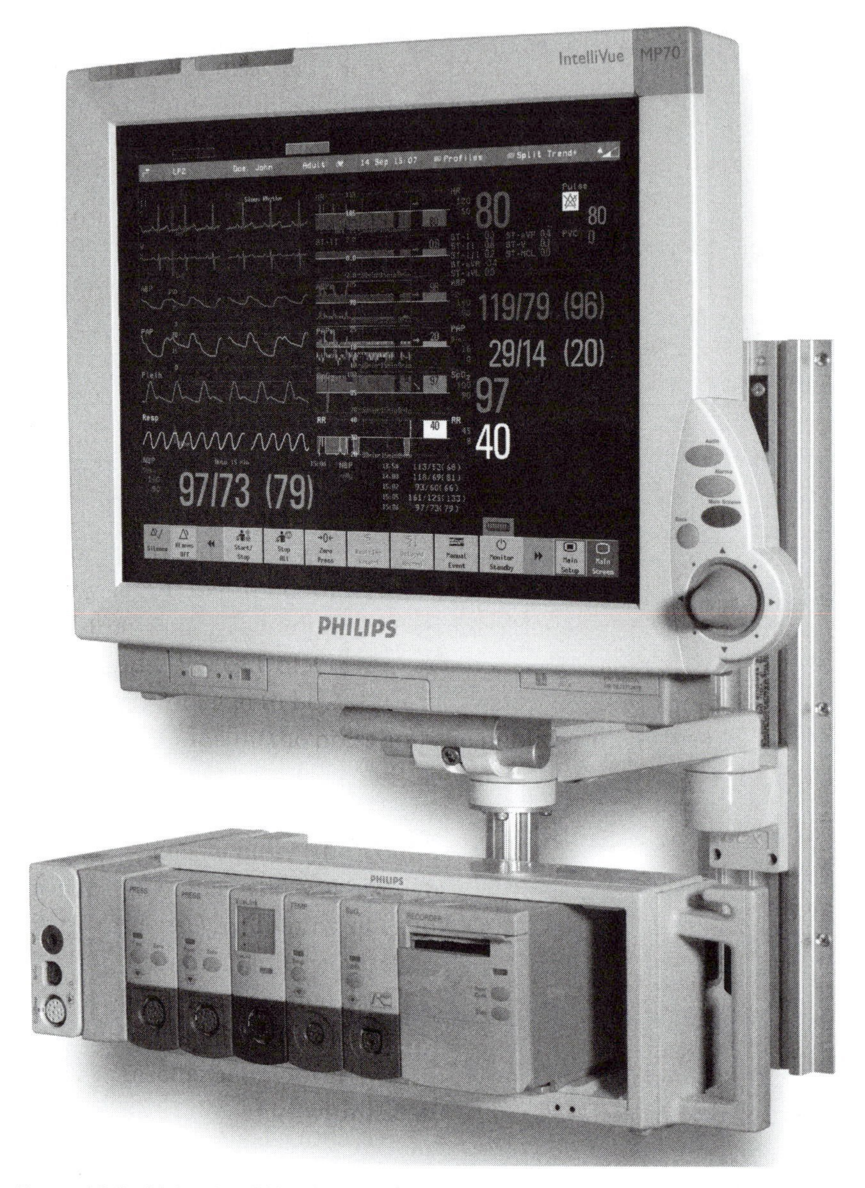

Figure 13.3. Philips IntelliVue Patient Care Monitor with modules for point-of-care testing. (Photo courtesy of Philips Healthcare.)

STUDY QUESTIONS

1. Describe the role of blood work and urine analysis in patient care. What are clinicians looking for? What types of information do they receive?
2. Describe the role of electrical measurements in automated blood tests.
3. Identify some commonly performed lab tests on blood.
4. Describe the type of information that is used to determine heredity (paternity, for example).
5. Identify the parts of a centrifuge. Describe their function.
6. Describe some of the specific precautions that must be taken by BMETs who work on centrifuges in order to prevent disease transmission. Include information about PPE from previous chapters.

FOR FUTURE EXPLORATION

1. Search the Internet to research the types of ion selective electrodes currently available. Document these types and identify their purpose. How have these innovations improved the speed and availability of a wide variety of patient tests?
2. Search the Internet to research, define, and describe a nephelometer. Identify the tests for which this device is used.
3. Research, define, and describe turbidimetry. How is this technique important in the processing of patient samples?
4. Research, define, and describe argose gel. Identify and describe its applications.
5. Research, define, and describe polymerase chain reaction (PCR). Describe how PCR is useful in DNA identification.

6. A common patient care monitor with point-of-care sample analysis is the Philips IntelliVue. Examine the product descriptions at http://www.medical.philips.com. Research, define, and describe laboratory tests the IntelliVue can perform at the bedside.
7. Barcoding technology is often used to process patient samples in the clinical laboratory. Identify and describe the method that is used to "read" barcodes. Identify and describe some common barcode formats.

14

Intravenous pumps and other pumps

LEARNING OBJECTIVES

1. identify and describe the function and purpose of an intravenous pump
2. identify and describe the function and purpose of a syringe pump
3. identify and describe the function and purpose of PCA devices
4. identify and describe the function and purpose of feeding pumps

Introduction

The use of technology to push fluids into a patient has evolved and expanded over time. Bottles of fluids used gravity flow to "drip" fluids into a patient. As technology expanded into health care, the bottles were replaced with plastic bags and gravity was no longer medically adequate. Electromechnical pumps are more reliable and provide constant monitoring of fluid delivery. In addition, the flexibility of some pumps has improved medication dosing and caloric delivery. Hospitals have come to depend on the use of technology to deliver fluids and medications to patients.

Infusion pumps

One of the most common pieces of equipment in the hospital, the infusion pump (also called intravenous pump, IVAC, and IV pump) delivers medication, blood, or fluid into the patient over a specific period of time at a particular rate. The name used may reflect brand names from the past, including IVAC. Many patients will be connected to several pumps delivering fluids to the patient through a vein (hence, the term "intravenous"). The pump ensures an accurate rate and, therefore, an exact dose. In addition, should there be a problem with the fluid delivery, an occlusion perhaps, or if the correct fluid amount has been delivered, an alarm will sound to alert staff. The electronic monitoring allows delivery of IV fluids with far less human intervention.

IV bags, filled with fluid, are connected through tubing to the pump, and then through the pump to the veins of the patient.

Settings on the device include delivery/flow rate as well as alarms to indicate blockages or empty fluid bags.

There are different techniques employed to pump in the fluid. One of the most common is a **peristaltic pump**. Rollers (in most cases) squeeze the tubing to push in the fluid. Traditional tubing from the fluid-filled bag to the patient is used.

Another method requires some additional **disposable cassettes** that hold the liquid for a piston to push the liquid. This

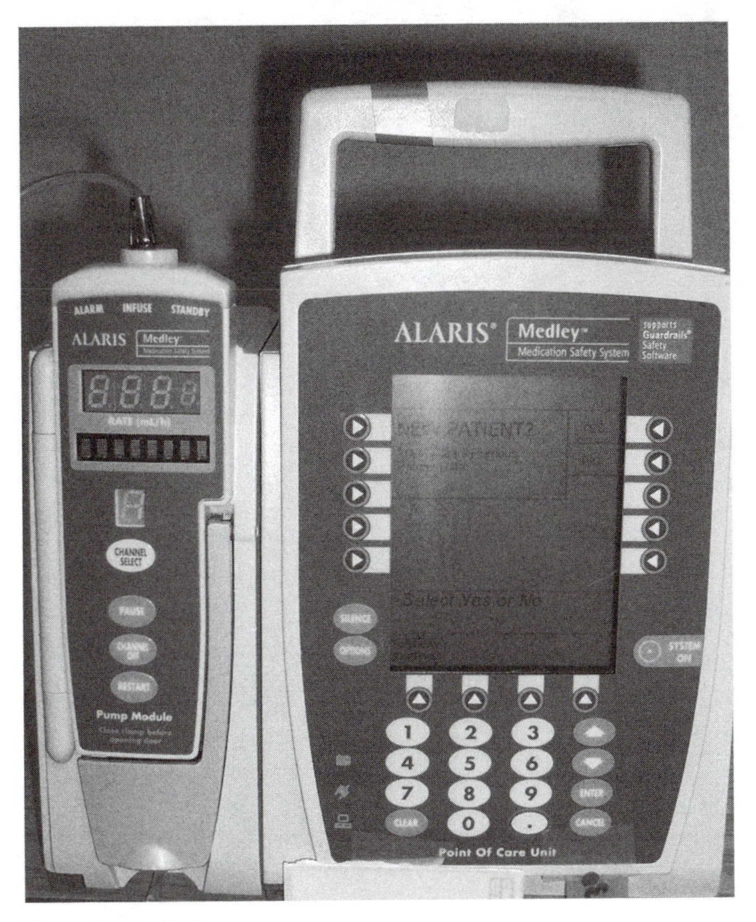

Figure 14.1. Alaris pump.

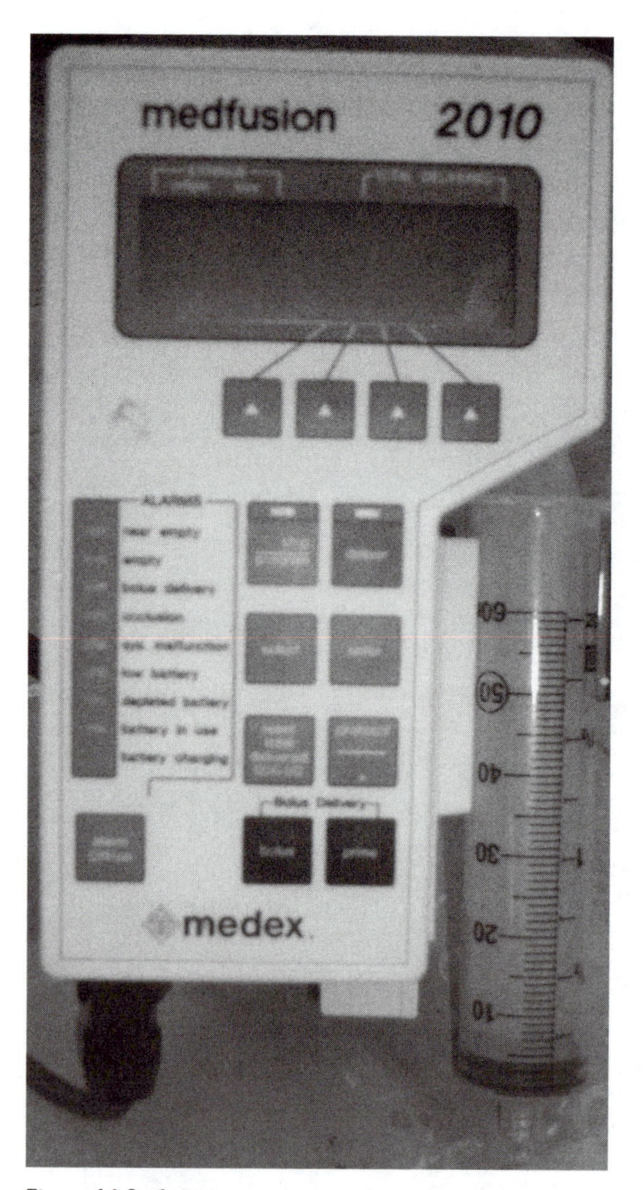

Figure 14.2. Syringe pump.

requires that the IV tubing be specifically chosen to work with the type of pump the hospital owns.

In Figure 14.1, a pump that has some memory for dosages of medications is shown. The blue screen in the center is the "brain," and the vertical unit on the left provides the pumping action to deliver fluids into the patient.

As a BMET, you will have to deal with many different manufacturers and models of pumps in your career. Many BMETs begin their internships working on these devices, which often take a great deal of abuse and are commonly in the BMET shop.

Syringe pumps

Syringe pumps deliver medication from a syringe by using a piston to drive the back of a standard syringe. This device is usually used for very small amounts of medication delivered very accurately. As with IV pumps, the medication is usually delivered to patients through their veins. Dosing instructions (time and amount) are programmed by the clinical staff. There are many manufacturers and models, but a typical syringe pump is shown in Figure 14.2.

PCA pumps

Patient-controlled analgesia (PCA) pumps deliver pain relief to the patient through an intravenous line, typically from a sealed cartridge, when a patient requests relief. Usually, the patient is given a push button to trigger the delivery of medication. Obviously, the pump limits the maximum amount of medication that a patient can receive. In addition, the pump can be programmed to limit the dose frequency. Generally, controls are located behind a panel inaccessible to the patient. Pumps with patient control offer better pain management and more accurate medication dosing.

Feeding pumps

Feeding pumps are used to feed patients suffering from chronic conditions who must be fed using a feeding tube. The patient is often connected to the feeding pump via a naso-gastric tube (discussed in Chapter 10). Because feeding liquids may be thick, gravity alone may not reliably deliver calories into the stomach. Feeding tubes are often connected to pumps that deliver enteral nutrition. This is especially true for premature infants who do not have a fully developed sucking reflex. A popular brand is the Kangaroo Pump.

Implantable insulin pumps

Diabetes is a very common condition in today's population. To deliver insulin, some patients use **implantable insulin pumps**. The name is a bit confusing because the pump is not implanted into the patient. However, the device does more closely match the pancreas by delivering insulin in a way that is much more continuous compared to injections that are typically self-delivered several times throughout the day. The patient has a port through the skin and a catheter is connected to a device that delivers insulin slowly. Some devices have the ability to monitor the patient's blood glucose levels and adjust insulin amounts as necessary.

STUDY QUESTIONS

1. Identify and describe the benefits of the use of technology to deliver medications and fluids.
2. Summarize the two methods of function of IV pumps.
3. Identify and describe the benefits of the use of PCA.
4. Describe what part of an implantable insulin pump is actually implanted.

FOR FURTHER EXPLORATION

1. Identify and describe why intravenous pumps are the most numerous device in hospitals. Why do many patients have multiple pumps? How does this impact BMETs who are responsible for technology support?
2. Intravenous pumps that use software to store medication types and dosing guidelines are becoming more common. Describe how these IV pumps can improve patient safety. How can these features dramatically increase the workload of BMETs?
3. Identify and describe some common types of fluids delivered by IV pumps.
4. IV pumps often stay with patients throughout their hospital stays. For example, a patient may move from an ICU to a regular patient floor. Describe how this may complicate equipment tracking. Research, define, and describe some asset

management techniques that can be employed to improve IV pump tracking.

5. "Smart" IV pumps were a big media event when introduced. They were seen as a significant way to decrease medication errors (one of the biggest problems in hospitals). However, there have been several large recalls with this type of pump. Explore the Internet for media announcements from the early 2000s that hail patient safety improvements. Research and document recalls from the FDA on Alaris pumps. Do the benefits seem to outweigh the risks?

15

Miscellaneous devices and topics

LEARNING OBJECTIVES

1. describe the role of BMETs in play therapy
2. define and describe fetal monitoring
3. define and describe nurse call systems
4. define and describe infant tracking systems
5. describe the role of BMETs in rehabilitation efforts
6. describe the role of BMETs in long-term patient care

Introduction

Many facets of the hospital culture engage technology as part of patient care, but they do not fit neatly into the traditional categories discussed so far. Children use technology (game consoles, for example) in play therapy. Women in labor are monitored using specific equipment. In addition, therapists and specialists, such as those involved in physical therapy and occupational therapy, also use technology. This chapter briefly examines patients who have long-term support needs.

Play therapy

Children's toys have increased in technological complexity over time. The role of play in a patient's recovery is well documented, and BMETs support this endeavor. BMETs ensure the devices are safe to use in the clinical setting, are secure from theft, can be cleaned appropriately, and are repaired when needed. In addition, some devices need adaptation to suit the needs of a particular patient. BMETs employed in children's hospitals face these challenges, which may call for creativity and sensitivity.

Fetal monitoring

Fetal monitors usually track uterine contractions and fetal heart rate simultaneously, often printing the information on long strips of paper. The medical team uses the two pieces of information together to evaluate fetal health and labor progress. A fetal monitor is shown in Figure 15.1. Chapter 5 describes fetal heart monitoring that occurs during labor. This monitoring can use indirect measurements (often using Doppler technology) or

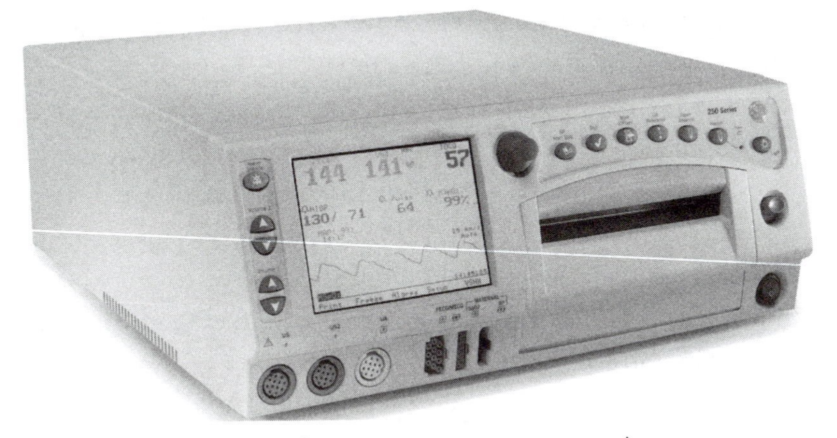

Figure 15.1. Fetal monitor. (Photo courtesy of GE Healthcare.)

direct, electrical measurements by the use of a scalp electrode, which connects to the unborn fetus. Doppler ECG monitoring uses a transducer placed on the woman's abdomen, usually held in place by an elastic band. Direct monitoring uses the scalp electrode and a reference electrode on the mother's skin. This monitoring method requires that the fetus be in a head-down presentation for delivery. In addition, the amniotic membranes that surround the fetus must not be intact.

Monitors used in labor can also measure the strength and duration of contractions. A pressure sensor (strain gauge) is placed on the mother's abdomen. The tightening movements of the abdominal muscles with each contraction are detected with the strain gauge. A medical term for contractions is "toco," and these transducers are sometimes called "toco" transducers.

Some fetal monitors use telemetry, which can transmit the physiological signals through the air wirelessly. This is very valuable for patients who may be encouraged to walk during labor. In addition, some monitors have transducers that are immersible for water births.

Fetal monitors are also used for tests on pregnant women, such as non-stress tests and stress tests. The results of these tests can provide the medical team with information about the fetal development and fetal stress. In non-stress tests, the heart rate of the baby is examined when fetal movement is detected. It is considered normal if heart rates increase after movement. Patients are typically provided a push button to record when a movement is felt so that the medical team can evaluate the fetal heart rate. Stress tests involve the comparison of heart rate to uterine contractions.

Using fetal ECG monitoring and contraction data together, medical staff can make evaluations of the fetus and its condition during labor. The collection of this data during labor can be printed on paper as well as electronically stored and networked. Medical staff located elsewhere in the hospital or at another location can have access to the electronic records to do "trend" analysis. These medical assessments compare the baby's heart rate with the pattern of contractions over time to assess the condition of both mother and baby. The ability to network data from laboring patients is a common hospital request.

There is a less-common method that can measure contraction strength directly with an intrauterine pressure (IUP) transducer. This is placed inside the uterus.

Nurse call systems

The technology that assists patients calling for assistance is often termed "nurse call" or one word "nursecall." What began as a button that, when pushed by a patient, turned on a light in the hallway, has evolved to include complex wireless paging systems linked to cell phones and pagers carried by medical staff. The devices are used at virtually every patient care bed and therefore many institutions employ one person (or a team) to focus only on this technology. Fundamentally, a patient in a bed is provided with a device that has one or more buttons that can be used to summon assistance, communicate with the unit front desk, and alert staff when there is a problem.

Infant Tracking

Infant tracking systems are used by hospitals to reduce the incidence of newborn abductions from the clinical setting. Tags worn by the baby, and in some models, the mother as well, vary in sophistication but, in general, can track the baby's location and sound alarms if the infant is removed from a designated area. Radio frequency identification devices (RFID) are often used to tag the infants. Antennas and portals are used to define the permitted areas and can trigger alarms when necessary. These systems can also be used to avoid infant switching, where a baby is sent home with the wrong parents. Generally, the tags sound an alarm if removed from the baby.

Rehabilitation

There are many staff involved in the treatments necessary to help patients regain a level of activity similar to that prior to illness or injury.

Occupational Therapist (OT): These staff members work to help patients regain or refine life skills such as tooth brushing or using an oven. Some people assume that OTs focus on employment skills. While this is possible, it is not the most common clinical application. BMETs are often asked to assist OTs to adapt everyday devices to make them safe for the hospital setting (adding a ground connection, for example) or appropriate for the needs of a particular patient. Some of the most interesting equipment requests BMETs may receive can come from OTs.

Physical Therapists (PT): These staff members assist patients with regaining or refining physical skills such as walking, standing, and movement in general. Many of their devices are purely mechanical in design. There are some devices used by PTs that can measure range of motion, some motorized exercise equipment, and ultrasound devices, which can soothe soft tissue injuries.

Speech/Audiology

Those involved in the evaluation of hearing use technology. Tone generators present audible information at specific frequencies and amplitudes and in variable patterns. In addition to common tone generators, the integration of lights and toys for young children is typical. For children who are too young to follow instructions, the combination of lights, sounds, and mechanical toys that move (the technician controls the on/off action) allows the evaluation of hearing.

Long-term care

Patients who have injuries or illness may be dependent on technology to support their organ functions. Some examples of long-term care support include ventilators, feeding pumps, pacemakers, and renal dialysis machines.

Renal dialysis machines replace the function of the kidneys by pumping blood through filters that remove the impurities that would normally be removed by the kidneys. Dialysis machines involve a great deal of mechanical devices and fluids that might seem more like plumbing. Technicians who work on dialysis equipment must receive specific training to ensure the work meets specific standards and ensures patient safety.

Prosthetics can be simple or complex, but they essentially replace the function of a body part. Examples include artificial limbs and Cochlear implants and could include artificial joints. While generally outside the scope of a BMET's responsibility, some technicians are deeply involved in the creation and support of these devices.

Terminally ill patients

Every human dies. How this happens and what role technology will play or not play becomes an ethical issue. Because there are several devices that can sustain life, "pulling the plug" is not a figure of speech. It is, however, a medical and ethical decision based on a number of factors, which have been identified as:

- Life-sustaining treatment simply delays death.
- Degree of physical or mental impairment is or will be so great that it is unreasonable to expect the patient to bear it.
- A particular treatment may be withdrawn or refused without regard to the medical opinion on its potential benefit.

To clarify a patient's wishes, two documents may be prepared:

- An advance directive contains instructions regarding health care decisions, especially in the case of incapacitation. It can include durable power of attorney and a living will.
- Do not resuscitate (DNR) order: A DNR is a patient's instruction not to restart a failed heartbeat or respiration. It does not mean that the patient will not be treated with medications. Patients who are DNR may still receive antibiotics and sedation or medications for pain. Do not resuscitate allows for a patient to die naturally if his or her respiratory or cardiac systems stop working.

STUDY QUESTIONS

1. Identify some children's toys that may need BMET support. Describe the support needed.
2. Discuss the prevalence of fetal monitors in hospitals that commonly deliver babies.
3. Discuss the prevalence of nurse call systems in hospitals. Given this information, discuss why many institutions have a dedicated person trained in the system.
4. Describe an example of a commonly used device that an occupational therapist might ask to be adapted for hospital use. Describe the skill the patient would relearn using this device.
5. List possible medical support devices needed by long-term disabled patients.

FOR FURTHER EXPLORATION

1. Read the January 2008 cover story in *24x7* about BMETs and play therapy at http://www.24x7mag.com/issues/articles/2008-01_01.asp. Summarize the relationship between the BMET, children's toys, and the support of patient play.
2. Research, document, and describe RFID systems used in infant protection systems (Hugs made by X-mark is a popular brand). How do the alarms or automatic door locks promote infant safety? Describe the potential concerns of parents regarding infant "tagging."
3. Research, define, and describe transcutaneous electrical nerve stimulators (TENS). How do they work? How do they benefit patients?
4. Hyrdrocollator is a brand of device that heats moist pain-relief pads. Describe and document how they work. How do they benefit patients?
5. Research, identify, and describe at least five major causes of permanent disabilities.
6. Google search "assistive devices disabled" and you will see the many, many devices available. Document how many are technically based (talking computers versus canes and walkers)? Describe a few of the technical ones. How has advancing age of the population expanded the availability of assistive devices?
7. Describe, document, and evaluate the ethics surrounding the use of experimental devices in terminally ill patients.
8. Research the **Uniform Determination of Death Act 1982,** which forms the legal basis for the recognition of brain death

in the United States. Research, document, and describe how brain death is determined.

9. Some court cases regarding end-of-life rights and technology become quite famous. Research and describe the end-of-life case of Karen Quinlan. How did the court eventually rule? What happened when her ventilator was disconnected?

Index